LEADERSHIP FROM THE DARKSIDE

THERE'S NOTHING MORE INSTRUCTIVE THAN A BAD EXAMPLE

Chris Kennedy

Chris Kennedy Publishing
Virginia Beach, VA

Chris Kennedy/Chris Kennedy Publishing
2052 Bierce Dr.
Virginia Beach, VA 23454
http://chriskennedypublishing.com/

Publisher's Note: This is a work of fiction. Names, characters, places, and incidents are a product of the author's imagination. Locales and public names are sometimes used for atmospheric purposes. Any resemblance to actual people, living or dead, or to businesses, companies, events, institutions, or locales is completely coincidental.

Ordering Information:
Quantity sales. Special discounts are available on quantity purchases by corporations, associations, and others. For details, contact the "Special Sales Department" at the address above.

Leadership from the Darkside/ Chris Kennedy. -- 1st ed.
ISBN 978-1942936503

As always, this book is for my wife and children. I would like to thank Linda, Beth and Dan, who took the time to critically read the work and make it better. Any mistakes that remain are my own. I would like to thank my mother, without whose steadfast belief in me, I would not be where I am today. Thank you.

"The greatest leader is not necessarily the one who does the greatest things. He is the one that gets the people to do the greatest things."

— *Ronald Reagan*

Foreword

It was my pleasure to be one of Commander Kennedy's first commanding officers, and to see him grow throughout the course of his career in the Navy. He arrived at Attack Squadron 34 as a brand new junior grade lieutenant; by the time he was done, he had commanded forces in every geographic region of the world. Over the course of his 20-year career he successfully led his aircrew into combat 37 times and brought them safely home every time. How did he do this? By blending them into a synergetic team where the whole was far greater than the sum of its pieces. Chris is successful because he is what every commanding officer looks for in their subordinates—he is a leader.

As a retired Navy admiral with 30 plus years of naval service, and more than 9 years in the corporate world, leadership is something near and dear to my heart, and it is something I have taught in US and foreign War Colleges, and spoken about on many occasions, both inside and outside the military. Although the application may vary slightly when you bring the military's 11 principles of leadership to the civilian world, the key tenets remain the same. *Leadership from the Darkside* is an innovative approach to discussing these principles; by showing where and how they were misapplied, future leaders are better able to understand why the principles are important and what happens when they aren't followed, and the story-based approach ensures readers will remember them a long time. The leadership principles in *Leadership from the Darkside* are solid and well-illustrated.

Chris Kennedy is an outstanding leader, and I am excited to introduce his book to you. I know you will enjoy reading *Leadership from the Darkside*, and that it will be help you become a better leader as a result. In a leadership landscape crowded with snake-oil salesmen, Chris Kennedy is a pioneer in his approach; this book is just as valuable to new leaders as it is to seasoned veterans. From intelligence officers who aren't, to a Marine officer who wants to dress like a bunny, the stories flow quickly and demonstrate the principles extraordinarily well.

RICHARD D. JASKOT
Rear Admiral, USN (Ret)

Introduction

When asked to describe his threshold test for obscenity, U.S. Supreme Court Justice Potter Stewart famously used the phrase, "I know it when I see it." Since then, that phrase has been used to categorize a variety of facts or qualities where the topic is subjective or lacks clearly-defined parameters. One of the hardest of these to describe is leadership. What makes good leadership? Is it a leader's ability to relate to those being led? Being able to share a vision for the future? Something else? Although people may know good leadership when they see it, the true nature of what makes a leader "good" is often elusive. What makes a leader "bad," though, is often far more obvious.

This book looks at various facets of leadership by analyzing stories of where leadership went wrong, sometimes horribly. There are stories from every branch of the military, and they are told from the viewpoints of both officers and enlisted. These stories, to the best of my knowledge, have been told without exaggeration; events portrayed actually happened, and in the manner they are related. Some may be hard to believe, but all are true.

The intent of this book is to demonstrate effective leadership techniques and illustrate what happens when they aren't appropriately applied, not to disparage any particular leader or leaders; the names have been changed to protect the identities of the people involved. In researching the stories, I found similar tales from different times and places, so while readers may think they recognize individual leaders, unfortunately, bad leadership is common enough that bad management examples (like history) often repeat themselves.

Contents

Chapter One
On Leadership

Good leadership is more important in the military than any other occupation. On any given day, leaders may have to order troops under their command to make the ultimate sacrifice for their country. Will that order be obeyed? In many cases, the answer depends on the qualities of the leader in question. If he or she can be trusted, the people being ordered are far more likely to make the sacrifice.

It's important to note that leadership is not a quality derived from filling a certain position, but is instead more about personality and behaviors. Having an executive position doesn't make someone a leader. It may make that person an administrator, but it doesn't necessarily make him or her a leader. Similarly, people who aren't in executive positions can still be leaders, and sometimes very good ones, often stepping up to fill the role of "leader" when their executives do not.

So…what is leadership?

Leadership is the colorless, odorless, flavorless gas necessary for an organization's well-being. You can't see it, smell it or taste it, but without it, the organization will most assuredly die. Without leaders, the organization and its members will wander aimlessly, not knowing where they are going or how they're going to get there.

Bad leadership in the military is almost a cliché these days; it seems every movie or book that deals with the military has officers or senior enlisted acting badly. That isn't representative of military leaders, though; in general, most are pretty good leaders, with quite a few being "great." They should be. The military spends an enormous amount of resources (in both time and money) training its personnel to be better leaders.

The military has to develop leaders. Its admirals and generals manage organizations with millions of personnel worldwide, who interact with millions of organizations across hundreds of countries on a daily basis. The American military would never have been as successful as it has without good leadership.

What does this mean to you? No matter what occupation you are in, your leaders can benefit from an analysis of the military's leadership program, and the organization will profit from applying the tenets relevant to your workplace.

How does the military teach leadership? By instilling 11 principles into its leaders, whether they are young enlisted troopers leading fire teams of soldiers or senior admirals in charge of carrier battle groups. These principles are the basis for good leadership and organizational success, regardless of the leader's environment or situation. If you want to develop good leaders, these are the principles you need to teach:

11 Principles of Military Leadership

1. Know yourself and seek self-improvement.

2. Be technically and tactically proficient.

3. Develop a sense of responsibility among your subordinates.

4. Make sound and timely decisions.

5. Set an example.

6. Know your people and look out for their welfare.

7. Keep your people informed.

8. Seek responsibility and take responsibility for your actions.

9. Ensure assigned tasks are understood, supervised and accomplished.

10. Train your people as a team.

11. Employ your team in accordance with its capabilities.

This book will analyze each of these principles, but it will do so from a different perspective than most leadership books. As was noted in the book's subtitle, there's nothing more instructive than a bad example. This book will address each of the military's leadership principles through the lens of several vignettes where leadership went wrong. The book will show how the leadership principle should have been applied, and it will make suggestions for how the principle can be applied in the civilian workplace.

Please remember, the intent of this book is to learn from past examples, not vilify the individuals mentioned.

Chapter Two
Know Yourself and Seek Self-Improvement

Over 1,500 years ago, the military strategist Sun Tzu theorized, "If you know the enemy and know yourself, you need not fear the result of a hundred battles. If you know yourself but not the enemy, for every victory gained you will also suffer a defeat. If you know neither the enemy nor yourself, you will succumb in every battle." If true, half the military battle revolves around knowing yourself.

Similarly, civilian leaders need to know themselves to be successful in today's workplace. If you know yourself, you can maximize your strengths and minimize your weaknesses. Unfortunately, it's not always easy to know yourself. Why? Because many times we don't see ourselves the way others do. Sometimes we are more critical of ourselves…but usually we aren't critical enough.

The television show "Survivor" is famous for bringing together a group of 'castaways' in a survival situation. Over the course of the show, the people have to band together for the good of the group, but they also have to vote someone off every three days. The show is a study of leadership in a fishbowl as everyone tries to be the leader (or at least assume control). The catch, however, is they need to do it in such a manner they don't appear to be the leader or to be in control because then they are a prime target for being voted off. During the show, the 'castaways' are periodically interviewed to see how they think they are doing; the positions in which they see themselves and their actual positions are often quite different.

So, how do you really get to know yourself? An honest evaluation to determine strengths and weaknesses is a good start, but usually more is needed to overcome the blind eye we tend to turn upon ourselves. Reaching out to your friends and superiors is the next logical step. Although friends may sugar coat the review, even if told not to, superiors will generally give you the information you're searching for as they are interested in your professional development.

The first vignette is a case where self-assessment or reaching out to peers would have helped the leader greatly. Sometimes, operating in austere conditions can be a source of camaraderie and teambuilding; other times, not so much...

Story #1. "Whose Sink is it, Anyway?"

"While on deployment in Afghanistan, everyone used the same shower tent; there were designated times for officers and enlisted, as well as males and females. One morning, the Operations Officer was shaving at one of the six sinks when the Commanding Officer (CO) walked into the shower tent. After greeting the CO, the Operations Officer continued shaving. At that point, the CO said, "That's my sink."

The Operations Officer, thinking it was a joke, replied, "Ha, ha, good one, Skipper."

The CO, unamused, replied, "No, really, that's my sink."

The Operations Officer finally realized the CO wasn't kidding, so he picked up his kit and moved to a different sink. The CO said nothing further; he just went to the recently-vacated sink and began to shave."

If you find yourself claiming territory like a five-year-old claiming a toy or putting yourself above and beyond subordinates in all matters (especially trivial ones), it's probably time for a little self-assessment. There was no reason for the CO to claim a specific sink as his own, other than to demonstrate he could. By making the Operations Officer move, the CO lost trust in the Operations Officer's eyes. If the CO would do something so trivial as to claim a sink as his own, what else would he do?

Another great time for self-assessment is when you find yourself doing things without putting any preparation into them. Think you know it all? Hopefully you're right, or things may not turn out as well as you hope…

Story #2. "Know What You're Going to Teach"

"My first experience with Navy leadership was during Plebe Summer (the summer before freshman year) at the U.S. Naval Academy (USNA). It has stuck with me ever since. We were being taught to march in platoon and company formations by the rising seniors and, as complete newcomers to the military, we had to have the maneuvers explained to us in detail. After the first explanation, we went out to practice, but it didn't seem to work. We tried hard but could never get it right. Something wasn't working, and we were told repeatedly it was our own incompetence. Before we went out the next day, we talked to one of the Marine Corps Gunnery Sergeants stationed at the USNA, and we learned the seniors had been teaching us the wrong techniques. This from guys who had presumably been marching for more than three years!

This episode lowered the seniors in our eyes and made us not nearly as willing to accept instruction and correction from them."

If leaders try to look like they know everything, and they don't, the leaders are going to lose the respect of their subordinates. What the story doesn't tell is how much abuse the seniors leveled on the freshmen while trying to correct them. It was probably quite a bit. I'm sure this made the loss of esteem far worse when the seniors were later found to be the ones doing it wrong.

So, how does a leader avoid this loss of respect? Take the time to read and study what you're going to teach and "practice before you preach."

Reading and study are great ways to assess your performance. Pick leaders you admire and learn more about them. As you read, try to focus on what worked for them, as well as the things that didn't. In most cases, you will find the leaders had a genuine interest in the people who worked for them (unlike Story #1).

Similarly, if there are areas you need to improve, develop a personal reading program to help you address them. For example, I had a supervisor at one stage of my career who was not what could be called a "people person." His personality was gruff, and he always gave the appearance (whether true or not) that he didn't care about you. Through some self-assessment, and I think a good dose of brow-beating from his wife, he realized he needed to be more approachable if he was going to become the leader he wanted to be.

He read and studied, and one day he was finally ready to try out his new approach. He had a task for me one Monday morning, and when he arrived at work he approached me and said, "Teddy, this week I want you to work on...No, wait a minute, I'm supposed to ask how you are and how your weekend was first." He paused and then asked, "So, how was your weekend?"

Now, obviously that was not the most tactful way to give the appearance he actually cared about me (and all the junior officers within earshot had a couple of laughs after he left); however, once we were done, we realized that he was at least trying, and we respected him for making the effort...even if his first try left quite a bit to be desired.

This attempt at self-improvement begs one of the long-debated questions in leadership: "Can leadership be taught?" Or, another way to frame the question might be, "Is leadership an art or a science?" To answer the second question, I would unequivocally answer, "Yes."

Leadership is both an art and a science. While some people may be born with more leadership talent than others (just like any other skill, whether intellectual or physical), I am a firm believer that leadership can be taught. People can become better leaders if they want to; otherwise, there would be no purpose for this book. Self-improvement is possible...but it has to start with an honest self-assessment.

Take, for example, a leader whose biggest problem is he can't control his temper. It wouldn't do him any good to work on minor issues, such as how he greeted his subordinates, if he didn't address his temper issue as well. While the "new and improved" leader might be able to greet his subordinates nicely, they would still be afraid to talk to him because they'll still be worried about him blowing up in their faces.

Even though you may not have as much natural leadership talent as someone else, it is possible to out-learn and outwork them, just like you can outwork someone who might naturally be a better baseball player. Others may have more natural leadership talent; if so, you have to outperform them through self-assessment, study and practice. And it all starts with the self-assessment...

Story #3. "Bad Leadership"

"Our first introduction to our new boss started with her saying, "I have favorites among the people who work for me, and if you are not one of my favorites then you will get nowhere with me."

The Information Warfare (IW) staff who supported the new head cryptologist was comprised of sharp sailors, seasoned chief petty officers, experienced limited duty officers and two lieutenant commanders who were nearing the end of their staff tours. In short, the IW staff in place to support her was extremely skilled. Our new boss only needed to provide the required guidance, and the IW team's experience, knowledge and work ethic would have ensured not only the success of the mission but her own personal success as well.

After a couple of weeks on board, however, it became evident her knowledge of the region was minimal, and her level of IW knowledge was similarly below expectations. This wasn't a showstopper; it could have been overcome if she had chosen to use the collective knowledge and experience of her staff to help get her up to speed. We knew our success rested on hers so it was in everyone's best interest to provide her with timely, relevant and accurate information so she could make the right decisions.

Unfortunately, she did not use us in this manner.

In short order, decisions were made without our counsel or awareness, with senior staff officers receiving erroneous information from her. This caused a tremendous amount of angst and confusion among the IW team. We wanted to do our jobs well, but we weren't given the chance.

Instead of communicating her needs or giving us direction so we could plan and develop solutions, we routinely received the following tasking from her, "I told the Intelligence Officer this, so make it happen." Worse, many times she didn't even tell us what she had signed us up for; we only found out when we were approached by other members of the staff who said, "Your boss told us you would do this for us." We found ourselves in a perpetual tail spin, seeking clarification on what was needed, managing unrealistic expectations and working ridiculous hours to keep up with the additional tasking.

To further complicate communications, our boss chose not to work in the facility we shared with the rest of the department; instead, she decided to work in Operations, a floor above us. This caused additional problems as it was not the appropriate environment to discuss classified material, and the space didn't have an appropriate way for her to communicate with us.

When faced with the realities that her promises to members of the staff could not be accomplished within the required time-lines (or accomplished at all, on some occasions), members of her staff received public beratings or admonishments (i.e., "I expected better of you," "You failed me" and "Who can I get to do your job?").

After several months, the senior leadership recognized that the loss of service they were experiencing was due to her lack of leadership and professionalism, and she was quickly reassigned."

Story #3 is a case study for "How not to be a good leader." In addition to not knowing herself, this officer was neither technically proficient (Chapter 3), nor did she keep her subordinates informed (Chapter 5). In fact, she was guilty of violating nearly all the tenets that will be discussed in this book. In her case, the biggest fault seems to be that there was no self-assessment made, which would have determined her need for self-improvement (nearly across the board). In Sun Tzu's eyes, she did not know herself, and she was destined to lose.

If you are placed in charge of a group, whether that is a military unit, a civilian production facility or a non-profit organization, you need to be the expert at what you're in charge of and know the people who work for you (e.g., their capabilities, personalities, and job descriptions). It's not your fault if you don't have the knowledge required to be successful when you are place there; it is, however, your fault if you don't recognize your ignorance and take the steps necessary to correct it.

Chapter Three
Be Technically and Tactically Proficient

Juniors have expectations of their leaders, like knowing what is required to accomplish the mission, and they expect their seniors to be capable of answering their questions. In short, juniors expect their seniors to be technically and tactically proficient, regardless of whether they are in a military or civilian organization. Just as junior attack pilots look to their seniors for guidance on how to fly their planes and bomb their targets, newly hired teachers will look to their mentors and the school's administration to help them become the kind of teachers they want to be.

In short, juniors expect to be able to go to their seniors for answers.

What does that mean to leaders? They need to understand that proficiency is expected of them; when they aren't technically and tactically proficient, they need to take the steps required to become so. Leaders need to be competent in their occupation as a whole and an expert in their specialty. As mentioned in the previous chapter, there's no disgrace in being placed in a position where you aren't the expert, because often it is the result of taking on greater responsibility (i.e., it's a *good* thing). For example, a leader might have been in charge of just a single division but was promoted to be responsible for a department made up of many divisions, and he or she might not have had any experience with them. There's no shame in being placed in charge of something you have no experience with, but there is also no excuse for continuing to be incompetent.

Many times, leaders in this position will seek to hide their incompetence, for fear of having their subordinates think less of them. In fact, the opposite is usually true. Being upfront about a lack of knowledge or experience and asking for help will show you care. In cases where your juniors know more about a topic, seek them out and get them to show you what they know. This makes them feel valued, and it builds their confidence in you when they see you trying to be the person the position requires. Don't just try to learn what is necessary, apply yourself to learn more; become the expert in the field. You will receive the respect of your juniors for making the effort. Besides, if you think you can hide your ignorance from your juniors, you're wrong. Trust me, they already know…

Story #4. "The Best Squadron on the Chalkboard"

"I was in an attack squadron during the war. On our first night of combat operations, our Commanding Officer (CO) went on the first mission, and our Executive Officer (XO) went on the second. They both came back, sat the junior officers down in the ready room and said (paraphrasing for the general audience), "It's not that hard; all you have to do is go out and do the jobs you trained for."

I'll never forget the CO saying something to the effect of, "I don't care if you plan a strike for five hours or five minutes. Either hit the target or don't come back, but if you're going to do that, send back my plane for someone else who *can* hit it." There was no "almost" or "close enough;" to the CO, you were either tactically proficient or you were not.

Coming back from the war, we got a new CO and a whole crop of new guys out of the training squadron. The new CO was *not* tactically proficient in the aircraft so he didn't focus on the results. He focused instead on the process, and only the planning was important; the outcome was irrelevant to him because his own skills were substandard. Every time a crew went out and missed a practice target, the skipper wanted to see their planning. If their preparation was good, they got a pass and nothing more was said.

The guys who had been in the war couldn't believe it.

I remember one night strike at a practice target. Two aircraft flew in for a routine type of attack. Nothing new, nothing hard; it was something we had all done lots of times. To make a long story short, it was uncoordinated, poorly timed and neither crew was even able to *find* the target, much less actually drop bombs on it. Their planning was good, though, so they got passes. This process continued, and I watched as the squadron lost the proficiency it once had…because it was now okay to fail.

My takeaway from watching this was that leadership is about results. Not "results at all cost or by whatever means," but about fostering an atmosphere of professional competence. It's about knowing the trade and being proficient at it, not just knowing "the tricks" of it. My last year in that squadron, we could have won a war against any enemy…as long as it was fought on the chalkboard."

Story #4 illustrates two very different kinds of leaders. The first commanding officer focused on the results and required all of his aviators to be technically and tactically proficient, and the second officer who was more focused on the process. Because he wasn't tactically proficient himself, the results weren't important.

While it might be nice to focus on doing everything right (and I'm not suggesting leaders should do things incorrectly or cut corners), at the end of the day, the only thing that matters to management or your investors *is* the results. The important question is, "Did you hit the target?" whether that target is an actual military target, a production quota to be achieved or an academic grade to be exceeded.

What do you do if you're not tactically proficient? Watch and learn from leaders who *are* proficient and "do what they do." Spend time with experts and learn from them, regardless of whether they are senior or junior to you. The most important thing is to be willing to seek improvement, and for many people that is hard. Another roadblock for some leaders is a refusal to seek information from their juniors (for a number of reasons, all bad), thereby losing out on key sources of information.

Leaders fail when they try to project an aura of competence they don't have or fail to seek improvement when they know they aren't experts. Still think your juniors won't notice your lack of proficiency? Too late; they already do, and it's going to cost you…

Story #5. "Prowling the Skies"

"Naval officers in general, and aviators in particular, will go to the ends of the earth and do nearly anything for a leader they respect. That respect derives from the junior officers' beliefs that the knowledge and experience of their leaders surpass their own; juniors expect their leaders to do their utmost to bring them home safely, despite the dangerous nature of their missions. Unfortunately, not all naval officers inspire that kind of loyalty. The CO of my last Prowler squadron failed in this regard.

This gentleman was often mercurial and petty. A former Marine Corps officer, he believed he could give any order, and the result would be quivering attention, instant obedience and enduring respect. However, the Navy is not the Marine Corps, and naval officers have different expectations than Marine officers. (Editor's note: knowing your organizational culture is important!) Naval aviators do not readily submit to being bullied by oppressive leaders if they are incompetent in the aircraft, tactically unwise or unsafe to fly with.

At the time of this story, the squadron was flying missions to enforce a No-Fly Zone. Returning from a six-hour mission in-country, the skipper was the flight lead of a two-aircraft formation as we approached our airbase. Normally, we preferred to enter the airport area under visual flight rules and execute a maneuver which gave us a quick downwind leg to a landing. That's how we flew at the aircraft carrier, and we tried to do it as much as possible even when we landed ashore so we could stay in practice.

As the flight neared the base, the weather started to close in significantly. While we were switching communications from our enroute controllers to the approach controller, the flight entered heavy clouds, and visibility was reduced to 50 feet or less. The pilot of the wing aircraft advised the CO he was intermittently losing sight of the CO's aircraft, and he said he intended to take separation and complete the flight as a single aircraft. However, the skipper wouldn't let him detach; the CO demanded the two Prowlers continue to make a section approach through what was quickly becoming zero visibility. The wing pilot couldn't maintain visual contact with the lead aircraft, which was only a few feet to the front and side of him flying at 250 knots. Still, the CO ordered the wingman to follow him, and then the CO stated the Prowler's radar could be used to maintain safe separation. This demand was made at high volume on the radio while flying in the clouds.

Those familiar with the airplane know the Prowler's radar is a fairly effective but ordinary ground mapping radar, and it is useless in any kind of air-to-air mode. It was sometimes possible to run the radar's tilt to horizontal and see a radar return from a nearby aircraft, but it was completely unsatisfactory for safely maintaining formation with another aircraft in zero visibility. This is especially true since the radar was in front of the navigator, well to the right of where the pilot could see it while he was trying to find the lead aircraft visually on his left. Still, the skipper asserted, again over the radio while flying in the clouds, that a competent aircrew could use the radar to follow his lead.

Using this technique had never been discussed in any of our squadron training sessions, it had never been demonstrated during fair weather flights prior to this incident and it had not been briefed before the flight. It was *not* considered a standardized flight procedure and was not mentioned in the aircraft's procedural manual. In fact, no one had ever heard of the technique before, and it would *never* have been considered a safe procedure for zero visibility formation flight. Still, the skipper expected the wingman to comply with his instructions.

Despite the CO's orders for the wingman to remain in formation and perform radar separation in the event of lost sight, the crew in the aircraft on the skipper's wing (wisely) disagreed. They followed accepted (and previously briefed) procedures and immediately called

"Lost sight!" and detached from the lead aircraft as they should have. They continued to the field under separate radar control and landed safely. To continue flying in thick clouds in the proximity of another jet could have resulted in the loss of eight airmen and two aircraft. The skipper made a demand that was tactically wrong, technically impossible and, most importantly, very dangerous.

Even though the crew of the wing aircraft had acted appropriately, the skipper was enraged. As the crews were debriefing in the ready room, he launched into a tirade, making allusions to deviant sexual preferences and disparaging their courage and manhood.

This sent up huge red flags throughout the ready room and the rest of the squadron. It exposed the skipper as technically incompetent, with an exceedingly poor understanding of the aircraft he flew. Most of all, it showed he was a dangerous pilot to fly with, the kind of pilot who would ignore published procedures and "make it up as he went along." Through his unsafe airmanship, the skipper provided all the justification necessary for the department heads and junior officers to totally disregard his leadership, in the air or on the ground. They began to think of him as an incompetent officer who had the one tag no one in Naval Aviation could abide: an unsafe pilot.

It is one thing to be an unsafe pilot in a single seat aircraft. The results of unsafe airmanship would most likely result in the demise of the guilty aviator. However, the Prowler carried a crew of four, all of whom were necessary to complete the mission. The CO's actions that day cost him the trust between the leader and the led, and he never recovered from that loss. It became difficult to get junior officers to fly with him; many asked to be excluded from his flights.

Normally, the Operations Department's flight schedulers distribute the less competent airmen with the squadron's best, in an attempt to mitigate perceived deficits in capability as well as to help improve the weaker aviators by exposing them to good examples. In this case, it became necessary to schedule more senior aircrew to fly with the skipper to ensure he didn't browbeat a junior officer into going along with an unsafe maneuver.

This anomaly in the flight schedule was put into place by the squadron's department heads by mutual consent, an act that clearly compromised their loyalty to him. It was the department heads who were charged with carrying out the CO's orders in the squadron, and it nearly led to anarchy as the department heads routinely began ignoring him. The boat had lost its rudder. This single incident led to a total loss of respect for the CO and minimized his control of the squadron. Fortunately, his command tour came to an end several months later, without any further incidents."

Could a similar loss of confidence occur in a civilian organization? Absolutely. I've seen it happen on several occasions. When leaders who aren't tactically proficient disregard the advice of their proficient juniors, they set themselves up for failure, not only at that time but also for the long-term due to loss of confidence.

There's no shame in being placed in charge of an organization where you aren't the expert, but there is also no excuse for continuing to be incompetent. There's also no excuse for not taking care of your people.

Chapter Four
Know Your Subordinates and Look Out For Their Welfare

The best leadership advice I received in my naval career was given to me shortly after I checked into my first squadron (thanks, Admiral Jaskot). The squadron's commanding officer took me aside and told me that if I only learned one thing about leadership, it was to take care of the people who worked for me. "If you do that," he said, "they will take care of you."

I said, "Yes, sir," and turned to walk away, like every young guy who thinks he knows it all, but he called me back over and said, "Think about it for a minute. If your people are worried about their wives, or their houses or their kids, they're not thinking about what they're doing. If nothing else, I don't want someone who is worried about his or her impending divorce working on my airplane."

Once again, I turned to walk away although I could at least see why that made sense. "It's more important than that, though," he continued, calling me back once more. "You are responsible for the people entrusted to you. You need to make sure they have the tools and training to do their jobs, to get promoted on time and to be successful in whatever they want to do. Your first duty is to ensure their success. If you do that, you won't have to worry about your own success because they will do everything possible to make sure you're successful, too."

Okay, now it meant something to me, so I asked what I was supposed to do. "Be involved with your folks. That way, you'll know what's important to them, and the things they need to be successful. When we get to the aircraft carrier, make sure they know (1) where they sleep, (2) where they eat, (3) where they work and (4) how to get from one place to the other."

All of a sudden, it dawned on me that this was a much bigger job than I thought. I didn't know my *own* way around the carrier. How was I going to help someone else? What did I know about all of these people who came from different places and different backgrounds?

The task seemed insurmountable, and I was bordering on being overwhelmed, but then he concluded with, "Be involved, ask questions and care about the people who work for you. It will all work out." I tried to follow that guidance during my 20-year career, and by and large, everything did work out pretty well. If you take care of the people who work for you, they *will* take care of you.

Oftentimes, though, it's more than that. You never know with whom you will end up working, so don't pass up a chance to, as the Boy Scouts say, "Do a good turn daily," and look out for the people who work alongside you as well. You never know where that will lead...

Story #6. "A Sister in Need"

"Our sister squadron had a plane go down for maintenance in Utapao, Thailand. The fix was a simple engine gauge change that required a technician and a quality assurance checker. It wasn't a difficult change; all told, it usually only took 45 minutes to complete.

My crew was repositioning back from Kadena, Japan, to the island of Diego Garcia in the Indian Ocean. Along the way, we had a two-day stop in Utapao so we got tasked to take the part and two maintenance technicians, both junior petty officers, from the sister squadron to Utapao. The problem started when we landed in Utapao with the crew, and no one from the sister squadron was there to meet their technicians. The two junior sailors borrowed our service ladder to gain entrance to their aircraft and then swapped out with one of their own.

As we were loading the vans to go to our hotels in Pattaya Beach, a favorite with sailors on liberty, my flight engineer (who was also my crew's senior enlisted) approached me and asked if we could wait for the sailors from the other squadron. When I asked where their detachment folks were, my flight engineer told me they weren't coming; the two sailors had been told to sleep in the plane, and the crew would be back around noon the next day to fly everyone back to Kadena.

I couldn't believe the squadron's leadership was willing to keep two-hard working sailors on an aircraft at one of the true garden spots of the world. One trip like this, done right, could erase months of hard, nonstop work, and the leadership of my sister squadron completely failed to comprehend it. They were missing out on a golden opportunity.

We waited, and my enlisted crew took care of the sister squadron's sailors that night, including supplying them with civilian cloths and taking them out for a night on the town in Pattaya Beach. We put them up with us in our hotel and got them back to their plane the next day in time for their flight back to Kadena. The more senior of the two went out of his way to thank my senior enlisted crew member. He was in awe that our aircrew would look out for them, and he asked how they could get a transfer to our squadron. For me, that one exchange, and the way these two gents were handled, confirmed the stories about bad morale and poor leadership in my sister squadron."

All it takes is a little sympathy and the ability to care about the people who work for you. In this case, it's obvious the sister squadron had leadership problems which were big enough for word to have gotten out to the other squadrons in the area. That kind of reputation encourages good personnel to avoid being transferred to the problem squadron, thereby ensuring a downward spiral of performance.

This translates similarly into civilian organizations, as word gets around an organization which units are the best at taking care of their people, and which ones are not. Taking care of your people becomes a vicious circle in reverse. Because your unit has a good reputation, it will attract the best people, which will help boost performance, increasing the unit's reputation, and so forth.

Not only is it important to look out for your personnel in their daily lives, but you also need to do what it takes to ensure their future successes. Subordinates need to be given the training required to get ready for promotion and career-enhancing opportunities to show their abilities. It's not just enough to say, "Yeah, you'll get that;" you have to follow through for them as well...

Story #7. "Why Plan When You Can React?"

"When I showed up at NAS Oceana to start my training, I had already been selected for the rank of Lieutenant Commander, having served previous tours in another aircraft. I was aware that making an aircraft transition was dangerous to my career, and that I might not get the promotion chances of someone who had started out in the community. Still, I wanted to make the attempt.

Soon after I reported to the training squadron, I visited the air wing's Readiness Officer to discuss my situation with him. My worry was that it would be more difficult to place me in a squadron than a first-tour junior officer since I would have no community experience but would have to fill a junior department head position. I would have to be carefully placed so I could get the experience I needed while still ensuring the squadron had the experienced flight leadership it needed. I was told there wouldn't be a problem; it would be easy to place me when the time came. I went back several times during my time in training, and I always received the same answer.

When I finished my training, though, the other members of my class got orders while I did not. When I went to Readiness to ask, I was told, "Well, it's not as easy to place you in a squadron because of your seniority and lack of community experience." Eventually, I got orders to a squadron just returning from cruise. There was no plan for my squadron to go on cruise any time soon, so I went back to Readiness to ask about the plan, because I knew I needed at-sea time in order to gain experience. Once again, I was told by Readiness it wouldn't be a problem.

After two years of relatively minimal sea time, the squadron was scheduled for decommissioning. I went back to Readiness to ask about orders to another fleet squadron, only to be told, "Well, you needed to have had more afloat experience by this time in order to be competitive, so there is really no place we can put you."

Ouch. Despite doing all the right things, and asking the right questions, this officer was suddenly unable to be placed; through no fault of his own (aside from switching airframes), his career was at an end. Nowhere during the process did any of his commanding officers or the air wing staff attempt to look out for him. Other people in the organization notice when things like this happen, and it leads to an organization-wide distrust of management. "If they did *that* to *him*, what are they going to do to *me*?"

In many organizations (the military included), the receipt of awards and rewards increases someone's opportunities for advancement. As a leader, you should ensure your juniors are recognized for their efforts; it is a tremendous morale builder. This is especially important in the military, as rewards are part of the criteria used for enlisted advancement...

Story #8. "I Got Mine, You Get Nothing"

"We had just returned from the war, where our squadron had been very successful. We put in several of our crews for awards, but later found out our squadron's awards had been turned down by the fleet's chief of staff because, "Your airplane isn't a combat aircraft; therefore, it isn't eligible for combat awards."

Well, I don't know about anyone else, but to me, nothing says "combat" like an enemy who's actively trying to kill you. When our aircraft was being chased by MiG-29's, it certainly felt like combat. When the enemy fired off an entire battery of SA-6 missiles (all 12) at my aircraft, it certainly felt like combat. I had enlisted sailors working for me who deserved those awards for putting their lives in harm's way, and the answer we received just wasn't good enough.

I asked my CO to request that the chief of staff reconsider the decision, as we *had* been chased by enemy fighters and we *had* been shot at by surface-to-air missiles, but the answer was "no." I begged him to readdress it and was told the answer was "no." I pushed the issue until the CO threatened to pull my wings if I brought it up again, but the answer was always, "no."

What did the CO ultimately do to resolve the situation? He put himself in for a Bronze Star, one of the highest personal awards you can receive in the military. He got his bronze star and was promoted; the folks who worked for him got nothing. He did not appear to care at all for the people who worked for him, damaging squadron morale and his reputation as a leader for the rest of his time as CO. Even the most junior sailor knew the CO couldn't be counted on to look out for anyone except himself."

While it is important to recognize subordinates' efforts, it's even more important to stay in touch with the mental attitude of the folks who work for you. If everything is going well at home and on the job, they will function at the highest level. When there are difficulties that intrude on their consciousness (as there always are), they will operate at a reduced efficiency. People going through severe emotional turmoil (divorce, death in the family) won't function at the desired level. In many occupations, it is *extremely* important to be on the lookout for these issues and give your subordinates time off to address them when needed.

Sometimes, though, you can't give your folks what they want. When that happens, you can at least address their concerns with a caring attitude...

Story #9. "Get Over It" Part I

"The strike group was nearing the successful completion of a six-month deployment. After three months in the Persian Gulf, the carrier was headed for home via a brief Christmas port visit to Western Australia. As the ship transited the Indian Ocean, the news was dominated by the possibility of an invasion of Iraq. The notion that our ship would have to go back to the Middle East was present and discussed, but the need to maintain a constant deployment and training schedule allayed our concerns about an extension of our deployment. We were also looking forward to, and more focused on, having several days of liberty in Western Australia. The carrier anchored a few days before Christmas, with leave and liberty set to expire the day after Christmas when the ship would head for home.

Unease took hold when the end date for liberty was extended two days. While every sailor will grouse for additional liberty to recognize their sacrifices, no one was naïve enough to think the chain of command was giving us additional time off for good behavior. A knot started in my own stomach, as this was the first tangible evidence "Big Navy" was in serious deliberations about turning us around.

Heading north these discussions intensified. As they did, so did the scuttlebutt among the crew. The opening pages of *The Red Badge of Courage* accurately capture the speed and ferocity with which a rumor takes hold of a military unit. The confines of a ship at sea only increased the intensity of the echo chamber. Leaving watch on New Year's Eve, I sensed our return would not be as originally planned. My sleep that night was interrupted by the rumble and shake of the ship changing course and picking up speed. I turned on the TV monitor and watched the compass spin until our course had reversed direction from due north to due south. Our trip home was over.

At reveille the next morning, the ship's CO spoke over the ship-wide intercom to inform the crew of what they already knew had happened. We had turned around and would not be going home. The captain laid out a simple, logical plan. The ship would first return to Perth for some needed maintenance. She would then transit to the Arabian Gulf, training enroute to re-establish the level of proficiency required of the ship's company and air wing. We would then, if so directed, fight the war. The ship would return to our homeport when it was no longer needed. The captain gave no dates – only general stages that, however nebulous, could be grasped and understood by the newest recruit and the saltiest chief alike.

Then the strike group commander took the microphone to talk to the crew. The admiral occupies an odd position in a naval force. He is resident on the flagship (so named because of the presence of the admiral's pennant or "flag" when he is embarked). Despite outranking the ship's captain, the admiral is not the commanding officer of either the ship or the air wing. Thus, a strike group commander has to maintain a delicate balance in allowing the captain to command his ship despite being both present and senior on the same vessel. All of this tension was present when the admiral took the microphone.

He talked for several minutes about the geopolitics of what was unfolding with regard to Iraq. It was a talk most of us wouldn't have remembered 10 minutes after he said, "that is all." That was, until he abruptly ended his description of the forces that had aligned to extend our deployment indefinitely. In that moment, his 30-plus years of experience as an officer and a leader somehow commanded him to instruct the 5,000 sailors and officers onboard that if they were in any way unhappy with the circumstances of not going home after five long months at sea, they needed to, "Get over it!"

The phrase physically hung in the air less than one second, but psychologically it suddenly filled the interior spaces of the ship as if it were smoke from an uncontrolled fire. "He did not just say that." "Are you f-ing kidding me?" "Get over it?" In that second, the most senior naval officer onboard intoned that all the sailors onboard were slackers. He was unable to comprehend the necessity to balance duty and country with empathy and respect, not only for the tasks that had recently been completed, but also for the unknown tasks which still lay ahead.

The gripes of the sailors about the possibility of a seven, eight or nine-month cruise were not anguished cries in the night of "why me," or "why not someone else?" Sailors often complain about additional work, and about others who have not worked as hard. These, however, are cast-off emotions of committed professionals, perhaps seeking acknowledgement of their sacrifice. Sailors' complaints are even a paradoxical means of communicating pride in their hard work and sacrifice. They are rarely to be taken at face value as evidence of shirking responsibility, duty or additional work. The sailors and officers knew that having recently completed three months in the Persian Gulf, they were the ones best prepared for the looming war.

Not only did the admiral's ill-chosen words denigrate the service of the 5,000, he drove a wedge between the ship's crew and the strike group staff. With complex combat operations looming on the horizon, the admiral injected cynicism and skepticism into the chain of command. He darkened the command relationship when it needed to be at its strongest. By barking at the crew to "get over it,' he ensured that they would *never* actually do so."

Here it is from another angle...

Story #10. "Get Over It" Part II

"It was dark in Combat at 0530 when I arrived to start turning over with the off-going Tactical Action Officer, or TAO, on New Year's Eve. Of course, it was always dark in Combat. That's the Combat Direction Center, occasionally called the CDC, but usually just "Combat." It didn't matter where we were on the planet, what the weather was like outside the aircraft carrier or what time of day it was. In Combat, the weather was always the same: dark, blue and cold.

The green lines on the link monitor gave me a representation of the real world worthy of the finest 1980's vintage video games, and I was immediately alarmed. My role for the shift was supposed to be making decisions about defending the ship as she passed through the Lombok Strait between Bali and Lombok, Indonesia, on our way home. The display did not show us just outside the strait. On the contrary, we were at least 60 nautical miles away. Worse, we were pointed south, away from it. Of course, the little green dots showing our track for the last hour or so formed a lazy, meandering circle, so maybe it wasn't as bad as it seemed…

"So, Goober," I asked the off-going TAO, "what's going on?"

"Oh, pretty much what it looks like," he answered. "We're just hanging out in the middle of the ocean."

Now don't get me wrong. I was completely *not* disappointed the strait passage wouldn't occur on my watch. A strait passage was one of the biggest pains faced by a TAO, especially in the world as it existed immediately following 9/11. You're in a 90,000-ton warship in restricted foreign waters, surrounded by vessels of every size and shape, dependent entirely on the vigilance of the lookouts to tell you what those vessels are up to, with minimal time to respond should one of them begin to behave erratically. When you add the realization that it's your fault if the Officer of the Deck maneuvers as you direct and the ship runs aground, what's not to like?

The problem was we were supposed to be on our way home. We were five-and-a-half months into what was scheduled to be a six-month deployment. Sure, the president was talking about going to war with Iraq, but we'd turned the Persian Gulf over to another carrier. We were off the hook…or so we had thought.

But here we were, turning lazy donuts in the ocean.

No sooner had I turned over and taken my place as the TAO, the phone started ringing off the hook. Everyone from the Officer of the Deck to sailors in my branch found some reason they "needed" to call Combat so they could get into a casual conversation with the TAO and hopefully figure out what was going on. It's not hard to notice there's no land in sight, and that you're obviously NOT engaged in a strait passage. My consistent answer was a helpful, "Hey, man, I just work here. They don't tell me shit."

Finally, after two hours of madness, the admiral came into Combat to use the ship's public address system, known as the 1MC.

He began auspiciously enough, "This is the admiral. I'm sure you have noticed we haven't begun our passage of the Lombok Strait. It has been decided more firepower may be required for operations against the country of Iraq, should the President order that. We have been asked not to proceed any further while we wait for orders from the Joint Chiefs of Staff, which I expect will be to return to be a part of the combat forces in the Persian Gulf."

Now would have been the time for inspirational words of encouragement and mission focus, or to sympathize with feelings of disappointment, or just to say, "thank you for your sacrifice."

Instead, he continued with, "I know some of you are disappointed we are not heading for home at this time…" (SOME of us?) "Well, get over it!"

I think he may have said more after that, but I don't remember any of it. I was trying too hard not to let my jaw drop open in visible astonishment.

After he left, I asked the Combat Watch Officer, "Did he really say 'get over it?'"

"He sure did…"

"Get over it" became an instant meme onboard, even though it was long before the days of memes…eventually, there were patches, T-shirts and all manner of memorabilia.

For reasons that could only be described as "fate" in a world where music could not be instantly downloaded, it turned out one of the twelve CDs for sale in the ship's store was "Hell Freezes Over" by The Eagles. Fans of the band may recall that album included the memorable single, "Get Over It."

We immediately went to work on the ship's executive officer, attempting to get him to use it as the ship's 'breakaway song,' the traditional tune a ship plays upon ending an underway replenishment operation. He'd have none of it... but when he went off the ship at an inopportune time, the operations officer, as his stand-in, approved it.

The ship went wild in miserable, defiant solidarity.

The admiral never recognized the tune."

It's not always going to be possible to give your subordinates everything they want, whether that's time off during a period of peak operations for the unit or that next big raise someone is due. When that happens, though, it's important to deliver the message with compassion and empathy if you want them to continue to function at high levels. Part of this is giving them information on why they can't have whatever it was they wanted.

Chapter Five
Keep Your Subordinates Informed

Do you want your subordinates to support you or to carry on in your absence? They need information in order to do so. Is there a big milestone coming up, or a high visibility event that's supposed to happen? If your subordinates don't know about it, they can't support it, nor will they know to give you the additional information they have which may be critical to the project's success.

In order for information transfer to occur, an organization needs to have a method in place for senior leaders to communicate with their subordinates. In many cases, this occurs at staff meetings of one form or another. The process will never work, though, if leaders don't care about communicating...

Story #11. "They Put the 'Fun' in Dysfunctional"

"I was temporarily assigned to a Marine infantry battalion conducting a humanitarian assistance mission. I had been assigned to the unit's higher headquarters, but as the mission was scaled back over time, the combatant commander decided to flatten the organization and eliminate the higher HQ. I was assigned to the next lower unit, the infantry battalion, to work in the unit's operations shop. It was here I observed the most sinister, diabolical and dysfunctional "command group" trio in my entire 22-year career.

To place the command climate in perspective, the command group "trio of misery" was the battalion's commander (CO), Executive Officer (XO) and Operations Officer—a lieutenant colonel, a major and a captain, respectively. They were failed leaders who created a maelstrom of stress, negativity and dread. Virtually everyone I encountered in the unit was miserable and driven to accomplish their mission not out of esprit de corps and pride, but out of simple fear of the consequences if they didn't.

As a relative outsider, I had the luxury of showing up, working my shift and disappearing again. My time with the unit was finite; I was assigned to augment the operations section until the unit redeployed, at which time I would then be reclaimed by the Joint Task Force and assigned elsewhere. I was not well-integrated into the unit, and for that, looking back, I am thankful. It seemed to me that the three constituent members of the command group were playing some sort of game where they tried to outperform each other in terms of being a jerk. Each time I observed one and decided, "He is the worst of the three," one of the other two would do something that completely recalibrated the poor leadership scale.

My best platform for observation of the command trio was at their daily staff meetings, where they would meet with the rest of the primary staff and the company commanders, including the commander of an attached US Army company, to discuss operations, support and planning. Usually, when a tyrannical commander is overbearing, unreasonably demanding or toxic, the XO serves as a shield to insulate the more junior commanders and staff members. If the CO and XO are both insufferable, the insulating function falls to the Operations Officer. The senior enlisted advisor, the Sergeant Major, has a limited ability to protect the junior officers but he should insulate the enlisted members from the commander's wrath. I do not remember who the Sergeant Major was, but he must not have been much of a presence if I do not even recall his name. Here, all three members of the command group were like-minded in their abusive leadership; there was no screening function.

The commander presided over meetings with a surly, angry demeanor. This was unjustified, because from my observation the unit performed fairly well in spite of its awful leadership. The XO took his cues from the commander—instead of setting priorities, encouraging collaboration and issuing guidance, he simply berated staff members in the meetings. The trio never once, to my recollection, recognized excellence or success. Staff meetings were not a time to exchange information, gain synergy and learn the perspectives of the other staff members – they were awful events that just had to be endured.

In the Marine Corps, general officers and senior colonels will often be heard saying that company command was the best time of their lives. The company commander of an infantry company has about 180 Marines under his supervision, and he is responsible for their welfare, training and employment. Company commanders are sufficiently junior that they get regular, meaningful contact with the lowest-ranking warfighters in the infantry, but sufficiently senior that they know their trade well and can conduct higher order planning, coordination and execution of operations with supporting arms and other units. It should be the best days of infantry officers' lives.

One company commander, who had been my instructor at the Basic School, and with whom I later developed a friendly, collegial relationship, told me the company commanders in this battalion—to a man—hated every minute of every day and could not wait to relinquish their commands and leave the unit. The Marine company commanders were largely cowed into submission, either by adherence to cultural service norms stating the senior person could act as he wished, or by concern over career-damaging retribution, but the Army company commander was not. I was friends with his XO, a first lieutenant, who often accompanied the company commander to these meetings. They were shocked a Marine unit could be this dysfunctional and poorly led, because the Marine Corps service culture purports to prize leadership above all else, and will announce this priority to anyone within listening distance. No information was ever passed at these meetings, and the Army officers often had little idea of what they were supposed to be doing to support the battalion.

I took away some leadership lessons from the two months I was in this unit.

First, I probably personally failed some Marines onto whom some of the fallout from this awful command group inevitably trickled. I do not know what I could have said or done to make it better, other than march into the Inspector General's (IG) office, lay it all out for him and ask him to either make some inquiries or initiate a command climate investigation. In retrospect, I probably should have. Approaching my immediate boss, the Operations Officer, would have been personal and professional suicide.

Second, healthy units need check valves like IGs, chaplains and senior enlisted advisors, and senior commanders should ensure their functions are unimpeded. This unit was an island unto itself, like in *Lord of the Flies*. Unfortunately, it never occurred to me in my youth to use one of these check valves to make the dysfunction in the unit known to its senior commanders.

Third, the unit's command group was an echo chamber. No good idea or input came from anywhere in the unit other than from the minds and mouths of these three tyrants. Afraid to speak, junior officers never asked questions or volunteered any information which might have helped the unit. The flow of information did not occur in either direction.

Fourth, left unaddressed, the failures of abysmal leaders like this can go unnoticed and unabated. The CO went on to become a full colonel and a very senior chief of staff in the Marine Corps. His reputation among his peers, I later learned, was that he was an irascible jerk, but I suppose he hid it from his general officer bosses well enough. The XO became a general officer. He may have excelled at colonel and general officer tasks, and he may have continued to the highest level of command, but he was an awful Major and an utter failure in his role as the XO. The Operations Officer faded into obscurity.

Fifth, I learned you can lead a unit through fear and intimidation for a while as a command tour is insufficient in duration to break the unit due to poor leadership. Twenty years ago, before the advent of social media which might have alerted others outside the command to the breadth and depth of the problems in this unit, it would have taken a personal act of intervention to break the cycle. In this case, the toxic leaders moved on, and the next commander inherited a train wreck; he spent his entire tour undoing the damage of his predecessors, while being held responsible for the conditions that were fomented on his predecessor's watch. This is a dangerous dynamic because it shows a toxic leader can hurt a unit not only for the period of time he is in charge, but also for years afterward, and can destroy the career of a competent successor whose only crime was to be unlucky enough to be sitting in the big chair when the unit fell apart.

Sixth, if not me, someone needed to stand up and sound the alarm—one of the company commanders, the unit chaplain, someone. The dysfunction in this unit was legendary, and it was plain for everyone to see, yet no one ever appealed to a higher authority to fix the problems in this unit. No one—this officer included—exercised the moral courage that would have been necessary to keep these poor leaders from infecting the hundreds of Marines under their charge. It is clear to me now, and I wish it had been clear to me then."

There is no way this unit could function effectively. Not only wasn't information making it to the troops who needed it, but there was also no way for the junior officers to pass information back to the command staff for fear of retribution. Deprived of the information they needed to make timely and effective decisions, the unit was sure to suffer. This was not only a short-term problem, but was an issue for many years to come, as was noted by the person who experienced it.

Just like a lack of knowledge can be devastating to a unit in both the short-term and the long, it can also have long-term effects for personnel beyond the time they spent in the unit. This may even occur when it isn't necessarily, "their fault..."

Story #12. "None of Us is as Dumb as All of Us"

"The Armed Forces have a collection of Staff and War Colleges that serve as academic centers for the teaching of the "Art and Science" of war at the mid-point of a service member's career. As an added benefit, they are often placed between two operational tours, where attending the school provides service members an opportunity to reconnect with their families and network with their peers.

My opportunity to attend one of these schools was in Leavenworth, Kansas at the Army Command and General Staff Officers Course in 2000. It was a great year, with endless lessons to be learned; most of the best ones were learned in the hallways between classes. This is one of them.

My class convened in the summer of 1999. Kosovo had just ended and 9/11 was two years, and in some ways a lifetime, away. The first semester of Army Staff College at that time was fairly regimented. Classes were 9-5, and there were "blocks" on leadership, tactics and logistics. These blocks were 4-6 weeks, where you studied one subject with the same instructors every day. The second semester in the spring was made up of electives chosen by the student, and the schedule ran like a regular college. You had between 10-20 hours of classroom requirements during the week; otherwise, your time was your own to study and prepare for class, unless there was a "special event" like a guest speaker.

Throughout the year, we had a series of distinguished visiting lecturers who would come to speak on one topic or another. These lecturers included all the service chiefs, a half-dozen currently serving and recently retired Combatant Commanders (generals and admirals who commanded geographic regions of the planet), and several combat heroes going back to the Vietnam era. Late in the fall, General Shinseki, who was then serving as the Army Chief of Staff, was scheduled to attend. Two days before his arrival, our instructors told us there was a requirement for all students to be at the school two hours early the next morning. When we asked questions about why we had to be there at 0530, we were told "I can't answer further questions; see you then," or words to that effect.

When we arrived the next morning, a questionnaire was handed out asking each Army officer to rate their recent experiences. There was also a question and answer section, and a comments section available at the end for amplification. We didn't know it at the time, but this was a fairly standardized instrument the Army had used for years to evaluate their officer corps at the mid-point of their service careers. Due to a catastrophic coincidence of scheduling, and a misunderstanding of how to administer the instrument, the Army was taken back by the responses it got.

Here's the problem. The instructors were told not to answer questions so they wouldn't bias the questionnaires. But to the students, it looked like the school was mining the student body to determine what their complaints were for the Chief of Staff just prior to his arrival, so he would not be shocked. Ironically, it was during a point in the curriculum focusing on leadership, and we had spent several weeks reading books like Lieutenant General H.R. McMaster's, "Dereliction of Duty," which centered on senior leadership's failure to speak the truth during Vietnam. Thus, there were 1,200 slightly miffed majors sitting in classrooms in Kansas at 0530, answering questions without being told why.

As an aside, the Navy, Marine, and Air Force Officers were turned away at the last minute and informed, "Oh, we didn't mean you guys." We found ourselves sitting in our cars for several hours because we weren't allowed in the classrooms, and it generally wasn't worth the effort to drive home and then turn around to come back. We were slightly miffed, too.

Following the collection of the questionnaires, life continued as usual. The Chief of Staff's lecture was interesting, and there were no inappropriate questions. About a month or two later, though, there was another oddity in scheduling. The Chief of Staff was returning, and this visit was for Army Officers Only. As a naval officer, I don't have firsthand knowledge, but as one might guess, the visit is all we talked about that week. From what I was told, the Chief of Staff welcomed everyone, pulled a chair onto the stage and announced, "Let's talk."

He had just gotten the results of the questionnaire, and apparently my classmates' morale, and more importantly their confidence in Army leadership, was falling off the left side of the bell curve. He explained he had just been out to visit, and we all seemed happy and asked good stuff, so what was the story? I'm sure he was relieved, but not amused, when he was told, "Uh, here's the thing sir, when we filled that out, we were all really pissed off."

To his everlasting credit, Shinseki, who took a day off from running the Army and flew halfway across the country for no real reason, then spent over an hour listening to that, and the inevitable, "But since you're here and asking, here's some stuff that bugs us." Some of it was probably true and relevant, some probably less so. Almost all of it went out the window two years later when the long war started.

If the story ended there, it would be amusing, but the other shoe dropped for me several years later. Two tours after my time in Leavenworth, I was chosen to serve as an Aide de Camp to an Army three-star general. This was in the days following the initial phase of

Iraq, and things were starting to turn south. During this tour, I learned there is a certain flow to generals' and admirals' annual schedules, where they spend most of the time doing their day-to-day jobs, but there are also annual or semi-annual points where they conduct conferences with all of their service or community peers. For the Army, the biggest of these is in October in Washington, DC, and it centers on the Association of the U.S. Army convention. This week is very much like a college reunion for the Army, and it includes many general officer meetings and symposiums.

During these meetings, the generals would go into a room, and for the next hour or two their aides, in crowds sometimes numbering in the hundreds, would try to catch up on unfinished tasks or talk amongst ourselves. I quickly learned a huge number of my fellow aides had been my classmates four or five years before, as we were all about the right seniority to be selected, and we had just finished three-year operational tours following Leavenworth. It was during one of these catch-up sessions I remarked on how many of us had been there together, when several others responded simultaneously, "Oh, we try to keep quiet that we were in the 'Whiner Class.'" Their response threw me, and I asked for further explanation.

One of my classmates who was present was serving as the aide for the Commander, U.S. Army Forces Command, the provider of operational ground forces to combatant commanders worldwide. He explained that following release of the questionnaire within the upper echelons of the Army, it was presumed our class was simply a "bin of bad apples;" hence, the "Whiner Class." Other aides present confirmed they had heard the anomaly referred to that way as well. One of the other aides had graduated the year before, and he confirmed that years later, the instructors, most of whom weren't even there at the time, still referred to our class as the "Whiner Class."

What I find fascinating about this incident is that when a phenomena is misunderstood or unexplainable, the reflex was to dismiss it as 1,200 bad apples, instead of rooting out the cause. If the results were legitimately that bad, it should have been a cause for concern and thus General Shinseki's return trip. There were three possible reasons for the results, 1) bad polling, which I think happened, 2) 1,200 bad officers, as the school administration, faculty, and senior officers in the Army argue, or 3) serious problems within the Army which nobody argued at the time (with the exception of their trouble-shooting skills). Of these three, only one conforms to Occam's Razor (the one with the fewest assumptions should be selected), and that choice is the Army's polling was bad.

Apparently the Army felt differently because, four years later and three years into the war, the common thinking appeared to be that for one year in Kansas, the Army assembled 1,200 majors of similar experiences, but dissimilar commissioning sources, year groups, career specialties, duty station records and circles of peers who were so unprofessional their responses invalidated a questionnaire developed over many years. Whatever bad apples had started the disease had been so effective they had successfully poisoned the entire class within three months of everyone's arrival.

Fortunately for the Army and the country, following graduation these officers would leave and be spread throughout staffs and operational commands, where they would be instantly rehabilitated. Three years later, they would conquer a country the size of California in 19 days...but they would still be dismissively referred to as a group as "Whiners." After all, that makes far more sense than making the Staff College review their statistical analysis procedures, and certainly more sense than introspection over a survey that showed eroding confidence in Army leadership."

An organization where the subordinates aren't informed is much like a ship without a rudder; it will wander aimlessly from place to place without purpose or goal, never achieving anything of note. Not only does the lack of information keep subordinates from knowing the goals of the organization (precluding their ability to help achieve them), it also robs them of the enthusiasm necessary to work hard on their accomplishment.

Information on topics like pay, promotion and benefits is especially important. Morale and esprit de corps improve when information is publicized on the unit's successes, and subordinates become more enthusiastic when they know they are making progress toward their goals. When those things don't happen, though, morale is sure to suffer...

Story #13. "Leadership and Perception"

"I was thrilled to learn I was chosen to command an Army Combat Engineer Company. I had just been promoted to the rank of captain when I received word I would be taking over command of a company in the 11th Engineer Battalion, which was located at Fort Stewart, GA.

Excited about the assignment, I spread the word to my colleagues and fellow engineer officers about my upcoming command opportunity. To my dismay, I received a great deal of negative feedback about the company I was taking command of and how bad a shape it was in; in fact, several of my peers and professional comrades warned me this particular assignment would be detrimental to my career.

Army officers' performance in company command, whether good or bad, is the primary determining factor for whether they will be promoted to major so I was more than a little worried by their negative comments. I decided to stop by the company, meet some of the unit's leaders and check things out for myself.

Upon entering the company orderly room, I was introduced to the Company First Sergeant, who was very positive, motivated and excited to meet me. After a brief tour of the company area (offices, Arms Room, Motor Pool, etc.), I met the company's executive officer and a couple of the platoon leaders. They all seemed to be highly motivated "sappers" (combat engineers) who enjoyed training and leading their soldiers. At that point, I couldn't see any problems with the company. The only person I hadn't met was the current CO, from whom I'd be taking command.

I went into the CO's office to introduce myself and found him sitting passively in the back corner. I stuck out my hand and introduced myself as the incoming commander. He weakly shook my hand and quietly said, "hello," without making eye contact with me. And with that first introduction, I discovered exactly what ailed the company.

The negative perception of this unit was due entirely to the void in leadership at the top of the unit, as exhibited by its company commander. As I saw firsthand, he had very little self-confidence, and I later learned he provided very little direction and vision for his subordinate leaders. He was reluctant to get involved, to take charge or to show any leadership or vision for the company and its soldiers. As such, the company often ran aimlessly while it tried to keep up with its sister companies in areas like training, maintenance and engineering skills.

The outgoing company CO gave little guidance to his leaders on what he wanted the unit to accomplish, and the information void went both ways. The subordinate leaders didn't have the information they needed to plan and execute their training and operations, and the CO was all too often absent from the unit's training. He had little idea where his unit stood with regard to achieving the standards required to prepare for deployment. Without a good understanding of his unit's current status, it was impossible for him to give orders for what to do next. He was in a vicious circle; lack of knowledge led to an inability to direct the subordinates' actions.

After I met the current company leadership and assessed the unit, I knew the job wasn't nearly as tough as it had been made out to be. The leaders and soldiers of the company were dying for some solid leadership, vision and clear guidance to help them achieve their objectives.

The platoon leaders responsible for planning the training and the platoon sergeants responsible for executing the training only needed to have measurable objectives and a clear understanding of what was to be achieved to turn the company around."

There was never any problem with the company's subordinate leaders and soldiers. As it turns out, they were highly motivated, eager soldiers who just lacked guidance from their commander. Once information began flowing from their next commander, their performance and morale soared. The negative perception of the unit was due entirely to the previous company commander's lack of leadership and communications skills.

Commanders set the tone for everything a military unit does, good or bad, and they are the unit's representative up the chain of command. They are the ones out front who represent their units. If commanders aren't confident, competent and do not have the mental courage and commitment to lead their soldiers, their units will fail.

Not only is it important to pass on operational information, but personal information as well. All too often, officers are great at passing on operational information vital to the unit's success, but they never pass on anything about policies and initiatives affecting pay, promotion or any of the troops' other benefits. One of the best training sessions I saw in my 20+ years of service was conducted by an admiral who took a large amount of his own time to develop and present a lecture on personal finance for the command's enlisted.

After the class, I could see it was an area routinely neglected by the people in charge, and one that the troops desperately needed (great job, Admiral Gallagher). The troops loved it and walked out with a better appreciation for personal finance, which had only been a mystery to many of them before then. Personally, I walked out thinking, "Chris, you have an MBA and more time available than the admiral. How come you never took the time to do this?" Periodic self-assessment, as discussed earlier, is imperative for improving yourself. Even when the answer is, "You suck."

Chapter Six
Ensure the Task Is Understood, Supervised and Accomplished

As any parent knows, if you want your children to do something, you usually have to tell them to do it (and frequently, especially for teenagers, you have to tell them more than once). As the parent is also aware, there is a huge difference between simply "doing it" and "doing it well;" that difference is often a matter of inspiration (through bribes, threats, etc.) and giving the children the amplifying information necessary to understand all aspects of the task.

Similarly, when leaders give their subordinates tasking, they need to do more than merely give the order to accomplish a task or perform a job. Leaders need to make sure their orders are clear and understood, and that the performance of the task is well-supervised. In the rush to get things accomplished, many times leaders don't listen to subordinates' questions or comments; their only guidance is to "just get it done."

And this can be a dangerous position to be in...

Story #14 "Flood the Drydock"

"I joined the crew of the submarine while it was in dry dock. It didn't take long for me to figure out the commanding officer. The boat had been in dry dock for most of the CO's tour, and he was chomping at the bit to get the boat underway so he could get some sea time on his record.

While sea time is great, and I am a Chief Petty Officer with plenty of it so I should know, the problem was the CO didn't fully understand everything being done to the boat. In his haste, he almost made an enormous mistake which could have seriously injured personnel and done tremendous damage to the boat. As it was, it only ended our two careers.

The critical event occurred near the end of our scheduled time in the dry dock. We were running behind schedule due to contractor delays which weren't our fault, but the CO didn't want to have to report we weren't going to make it out on time. He was determined we would leave the dry dock "on schedule"...whatever that took.

We worked hard and were very close to being complete on the scheduled date; however, we still weren't quite done. When the CO asked if we were ready to flood the dry dock to get underway, I had to tell him, "no." He flew into a rage and said we'd just finish whatever was needed at sea; he wasn't going to be late.

I tried to tell him we weren't done with the work being performed, and what that entailed, but he refused to listen. He went ashore and told the port facilities folks to flood the dry dock. I tried to stop them, but they were more afraid of him than they were of me.

Water began flowing.

The problem was that some of the exterior valves on the submarine were open and couldn't be shut without hydraulic power, which we didn't have yet. There was nothing to stop the water from rushing into the boat when it reached the open ports. Put simply, the boat was going to flood and ultimately sink at the pier. It was possible people might be trapped and drown; it was certain a large amount of equipment *was* going to be destroyed.

I tried to talk with the CO, but he wouldn't listen to me, so I called the base CO. He listened, and then he called the facilities folks, stopping the dry dock flooding just before it reached the level where it would have entered the submarine.

The base CO was furious, the boat's CO was furious and I was furious. As the junior person, I lost, even though I was right. The only (small) consolation was the boat's CO also lost, and he was shuffled off to a desk job somewhere. All of this was unnecessary. If he would have just let me explain the nature of the job that still had to be completed, none of this would have happened."

In many cases, how you do something is just as important as what it is you are doing. Regardless of what else is going on, a leader needs to take the time to ensure the task is not only understood, but also any reasons why something "out of the ordinary" is necessary. Although it may (infrequently) be necessary to say "Just do it," or "I order you to do it," for operational or safety reasons, it is usually better if you can give subordinates some context for why you are doing something outside of standard operating procedures. This gives subordinates the information they need not only to (1) accomplish the task, but also to (2) find a different method that might accomplish the stated goal more effectively.

In my 20 years of naval service, I think I only gave three "orders," including the last one where I ordered my crew NOT to go throw something into the lava flow of a nearby volcano. It was the last night before a war was about to start, and they had heard it was good luck to throw something into the lava. In my judgment, it was even better luck to stay far away from the molten lava. I didn't know how it would go if I had to call back to base and say, "Skipper, can you please send us a new radar operator? Ours fell into some molten lava trying to wish for good luck." I didn't know...but I couldn't see any way it would have gone well. Certainly, my leadership would have been called into question, and justifiably so.

The problem with saying "just do it" is that when you fall back on ordering people to do things from a position of authority, you have probably closed off all avenues for getting feedback. You had better be sure that the decision is correct and necessary, as there won't be any further information flowing back to you...

Story #15 "The XO Enraged"

"Who, What, When, Where and Why comprise the "5 Ws" of any operations order; therefore, when I asked why we needed to move, I was greatly surprised when the battalion executive officer (XO) answered with one of the most expletive-laden responses I have ever heard. It went something like this (edited for polite company): "Did you just ask bleeping why?! I'll tell you bleeping why! Because the bleeping colonel said to bleeping move! You are nothing but a bleeping captain—you don't bleeping ask bleeping why, you just bleeping do what the bleeping colonel or the bleeping major says! Do you bleeping understand?! If you don't bleeping understand I will bleeping fire your bleep and bleeping find someone who doesn't bleeping ask bleeping questions!" In my memory the tirade went on a lot longer than it takes to read this, so either my memory is faulty or he repeated himself a bit—either option is equally valid. All that to a simple question.

The "Why" in an operations order is there to establish the purpose or need. As a new battery commander I thought I needed to know why we were moving. Is there enemy activity in the area? Is there a herd of local fauna moving through? Is there a storm brewing that is causing us to evacuate? "To practice moving the tactical operations center (TOC)" would even be a perfectly valid answer. I was looking for something that would constrain my options while moving; for instance, enemy activity in the area would likely limit my options for movement by making certain paths impassable. Similarly, a herd of oryx might limit movement, as soft-sided Humvees are no protection from the long horns on large, angry antelope. A flashflood warning would have made us avoid low lying areas.

We were in the field for a major air defense exercise in the deserts of Texas and New Mexico, and our unit was providing short range air defense to maneuver units. We met our movement timelines and beat the supported unit into the field by a couple of days. An Opposition Force (OPFOR) was involved, but we knew nothing of their movement timetables so I felt my question of "Why?" had validity. Instead of constraints I received curses. Not being immune to such treatment, I have to admit to losing my own temper. Thankfully, I didn't say anything to the XO I would later regret. I did come to the summary conclusion we were simply moving the TOC for practice.

For full disclosure and in the XO's defense, he was under a lot of pressure. One of our line batteries was collocated with the battalion headquarters and headquarters battery (HHB), at a location either selected by him or recommended by him and the staff. Anyone familiar with the desert in that part of Texas knows it is an ancient lake bed, which floods when it (infrequently) rains, which happened a day or two after we got into the field. When the HHB moved to escape the flood, we lost communications with them. Most of the night was apparently spent out of communications, and the soldier assigned to maintain our battery communications fell asleep with the headset on his head and failed to wake his relief...and never heard the HHB calls when they finally came. So the XO was already highly stressed and unhappy, and he used me to release some of his anger.

As an individual who was high-tempered and low-tongued, this reaction is understandable, I suppose. However, I was "raised" to understand that officers were to behave as gentlemen. Such an outburst was inappropriate and cursing any soldier, much less a fellow officer, even more so.

A good officer can certainly show passion, but must remain professional. This includes maintaining control of one's tongue even when angry, and using words that show proof of a larger vocabulary than is traditionally found on a seedy dock. There are two overarching reasons for doing so: first, it is the proper way to behave; second, the troops expect officers to behave so. It is simply the way the game is played.

Such outbursts impact how others perceive you for the rest of your career. Everyone who witnesses events like this remembers them...and they tell others. Now, twenty years after that event, the former XO's temper has mellowed, and he has even learned to curb his tongue somewhat, but there will always be reservations in my mind when dealing with him. If I feel that way after only dealing with him occasionally back then, consider how those who worked closely with him feel.

The lesson learned here is to act professionally regardless of the circumstance. Act the part you are playing even if you don't feel like doing so. This will reinforce your subordinates' opinions of you as an officer, and build their trust in your ability to lead. Yelling at your subordinates shows you cannot lead yourself when stressed, and if you cannot lead yourself when the going gets tough, how can they expect you to successfully lead them?"

The advice given is as valid in the civilian world as it is in the military. If you ask a question when told to do something, and your boss jumps down your throat, the next time you're given a task you're probably going to just do it, even if you know it's incorrect, rather than ask for clarifying instructions. Doing something you know is wrong is *not* the optimal response; not only is it erroneous, but your subordinates now see you doing something incorrect, and they begin to question *your* judgment, as well. Not being approachable has long-lasting and long-reaching effects across the entire organization and the operations being conducted.

Finally, as we will also see in the "Make Sound and Timely Judgments" section, if you can't justify a tasking you're giving to a subordinate, it's probably not worth doing in the first place...

Story #16 "Well, Somebody's Got to Take the Fall"

"Sometimes you don't just have one bad leader. Sometimes you have a pack of them.

In my last job, I was an analyst for a military service-affiliated organization that, among other things, designed and executed analytic workshops. One time I was called in late in the process to participate in an event we were hosting for someone else. (Note: Hosting someone else's event was something we officially Did Not Do, but none of my colleagues could remember the phrase "we don't do that" being followed by "...and so we won't do it now." "We don't do that" was always used as an introduction to an explanation for why this was the exception.)

The topic for the workshop was an organizational structure one of the military services was experimenting with, and I will admit I didn't really understand it. "That's okay," the departmental leader told me, "you don't have to do the analysis. You just have to read the contractor's report and give us feedback on the general quality of it." I can do that, I thought.

I should have known better.

I joined the experiment during its final design meetings, and I discovered something I learned had been true from the beginning: the sponsor knew the answer he wanted, and the contractor was determined to give it to him. Our lead design person, who was there primarily as a "wedding planner" (remember, we were just the hosts) was being consistently outvoted. The event was really just a smokescreen to make it look like real analysis had taken place, and worse, our organization's name would lend validity to the results. This, in a nutshell, is why our official policy was not to host other people's events, and the reason why the worker bees had argued we shouldn't host this one.

Sometimes, as a worker bee, you are told "you're opinion has been noted, now shut up and salute." We did so, and that was that. Until late Wednesday afternoon during the workshop, that is. At that point our departmental leader called me in and said "listen, we're going to need you to write a report."

"I can't write a report," I said. "I was told NOT to do any of the things I would need to do to write a report. There is no Data Collection and Analysis Plan (DCAP), we haven't been capturing electronic worksheets or briefs and we don't have trained data collectors in place."

"Don't care," came the reply. "This has turned into a political issue. The administration is concerned we're hosting someone else's event (!), so we have to write a report." The workshop had started Monday at noon with intro briefs; Friday morning were the out-briefs. So the meat of the workshop was Tuesday, Wednesday and Thursday, and we're having this conversation at 4:00 p.m. on Wednesday. The workshop is already 2/3rds over, and I am being told to pull a data collection and analysis effort out of thin air.

Now, as it turned out, I wasn't totally hosed. I had asked some friends who worked this issue for a sister organization if they wanted to attend the event. The idea was they could give me a thumbnail sketch of what happened in the various cells, allowing me to be a more informed reader of the contractor's report. They had been taking notes and, while it wasn't the same as a fully instrumented event with a DCAP, it was better than nothing.

After the workshop concluded, I sat down with our lead design person, the guy who was consistently outvoted by the sponsor and the contractor. I explained my plan: a brutally honest report that would outline all the issues with the event, exactly as it happened. "Call a spade a spade" and all of that, emphasizing YET AGAIN why we don't host other people's events.

This idea did not make my colleague happy, because he remembered something I did not: as our event lead, he was going to take the blame, whether or not it was his fault. Unfair? Of course it was, but neither our departmental leader nor the administration over him was above assigning blame, despite the fact they had both explicitly over-ruled our objections and had refused to give the authority necessary to do the event properly. Heck, it couldn't be THEIR fault, now could it? So it had to be my colleague's fault. I had been there long enough to see he was probably right, or at least wasn't being paranoid.

So, I'm faced with writing a game report that is not a lie, while not actually the truth, either. I'm perversely proud of the final result. The conclusions section largely relied on reproductions of the slides the sponsor put together (before the event, I think) for the outbrief, showing how the workshop results fully supported his position. A careful reading will show the only actual conclusion of the report was that the sponsor liked the event.

The leadership exemplars don't end there, however, and I am getting a bit ahead of myself. As I said, this report had become something of a hot potato and was only possible because I had called in some friends for help. Three of the four gave me quick write-ups within a day or two. The last one asked for a bit more time. She was doing a report for her department anyway, and she would just give me that. "No problem," I said. "Send it to me by close of business next Friday," which would give her a week and be about two weeks after the event ended.

The next week was annoying. "Where's the report? Where's the report?" Jeez, there's no way I would normally have had the report done that fast (doing analysis takes time), but the same political pressures which led to the necessity of a "cover your butt" report in the first place were also demanding it be done NOW. I called my friend, but her voicemail said she was out of the office that week, and I vaguely remembered her saying something about being out of town. I was still confident she would get it to me on time.

Things finally came to a head on Thursday, when the deputy department head told me that I couldn't go to an offsite meeting with a potential business partner, "because your work's not done." I explained that (1) I am waiting on information from one of my data collectors, and (2) it's on schedule for a Monday delivery, as promised, because I plan on coming in over the weekend to finish it up. Hence, going to my Friday meeting didn't actually affect the report delivery date either way – it could be done late Friday or late Sunday, but no one would see it until Monday morning, anyway.

"You can't go until it is done," he responded. Apparently I was in the doghouse because I was trying to do a good job and actually use data to do the analysis. (I'll note that the deputy department head came into the job a year previously with ZERO experience in any of this. As far as he was concerned, analysts just "Made Stuff Up," and hence, he couldn't understand why I wasn't done three hours after the event ended.) Well, an order is an order; I've told the boss I'm waiting on data, and he says he doesn't care. Par for the course.

So, I stayed late Thursday, came in early on Friday, and sent it off to the department head at about 7:00 p.m., as instructed. I didn't get to go to the meeting, despite being one of the two critical people needed for it. The woman on whose input I was waiting was so pissed I launched the report without her that she never spoke to me again. (I pointed out this was actually a good thing for her and her department, as she wouldn't have to explain why my report didn't adopt her "the event was a farce" conclusion, but she took it as a personal betrayal that my boss ordered me not to wait.) I didn't get any feedback on my report from anyone above me in the chain of command, so I assumed this meant I sufficiently whitewashed the whole affair. That would have been fine, if that's where it ended.

It didn't.

Six months later, the deputy department head sent me an email: "Hey, where's that report? It's really late, and the administration just asked about it." I checked my files (ALWAYS save a complete email record!), and reported back the exact date and time I sent it to the department head. As it turned out, the report was SO CRITICAL the department head immediately lost it in his inbox, the administration forgot about it and the deputy never followed up. To the best of my knowledge, the design lead and I are the only people who ever read it."

Poor tasking with abysmal instructions and no follow-up. Several relationships wrecked, and a business deal tanked, all for a report that no one ever read. Nice.

Chapter Seven
Train Your Unit as a Team

When I first came into the Navy, I flew the A-6E Intruder aircraft. The Navy's premier bomber off the aircraft carrier, it was a two-person aircraft, where the pilot and bombardier/ navigator (B/N) sat side-by-side. It was very much a "crew concept" aircraft, as the pilot flew the aircraft while the B/N (me) operated the communications, navigation and weapons systems. Both of us needed each other if we were going to successfully accomplish the mission.

To that end, a pilot and a B/N would usually fly together for six months or so. During that time, each came to know what information the other needed and, most importantly, when they needed it. After a few flights together, a synergistic effect took hold; the team was much more than the sum of its parts.

That didn't mean we were both perfect, but we knew each other's strengths and weaknesses, and we knew how to help our crewmate when he needed it. Training as a team made that synergy possible.

Teamwork makes every organization function better, whether military or civilian (as can be seen in most professional sports teams). It starts with knowing your job first, but also relies on knowing the duties and responsibilities of your co-workers so you can fill in when/where necessary. Whether assistance is necessary due to an increased personal workload or because of your teammate's illness, being able to cover for them when needed helps the organization function better and become more effective. For teamwork to occur, though, the leader has to set the conditions and organizational climate in which teamwork can grow and thrive…

Story #17. "Power Play"

"Shortly after our new CO took command, it was time for our first department head meeting with him. At 0900 on Monday morning, all the department heads were assembled, along with the XO. As the Administrative Officer, it was my job to inform the CO when everyone had arrived, and the meeting was ready to begin. At 0900, I informed the CO that everyone was assembled and ready, and I returned to the conference room. By 0910, we all began to wonder if the CO had gotten tied up with a phone call or some other squadron business. At 0915, I went back to the CO's office and once again informed him that everyone was present, and we were ready to begin. His terse response to me was, "The meeting will begin when I get there." I replied, "Aye, aye," and returned to the conference room. We waited, and then we waited some more. Finally, the CO came in shortly before 1100 and informed all of us that *he* was the CO, and *his* meetings would begin when *he* was ready. There had been no rhyme or reason to the almost two hour delay, other than to prove to us that he held the power, and we were at his beck and call."

In an organization of professionals, this display was unnecessary. I wasn't there, but I can guarantee all the department heads, and the XO as well, knew the new CO was in charge without having their noses rubbed in it. Over 20 man-hours were wasted so the CO could demonstrate he was not a member of the team; in fact, he was far above it. This sets an example for the department heads to implement the same policy of empire building within each of their departments, and so on down the chain of command.

A divisive display like this will not help the members of a civilian organization function as a team either; it will do just the opposite. The members of the team should be focused on accomplishing the organization's mission, not carving out individual fiefdoms from within it. The squadron would have been better served to have the CO at the meeting on time, with an opening statement on the mission of the squadron, and how they could all work together to achieve it. That would encourage teamwork, the same as it would in a civilian organization.

Oftentimes, encouraging participation in recreational or company events is a way to help grow teamwork within an organization. It lets your subordinates see you outside the confines of the organization, building trust in you as a person. Participation in company events can be a positive way to increase teamwork…if done well…

Story #18. "Mandatory Fun…and Beverages"

"A friend of mine was stationed on a staff based overseas. As sometimes happens when tour lengths are fairly short, his XO didn't bring his family with him when he moved; instead, he left them at his previous duty station so his children could continue in the same schools without interruption.

While that worked out well for the XO, many of the officers on the staff had chosen to bring their spouses and children with them in order to experience the culture of the country. Neither of these were problems in and of themselves.

But the XO liked to drink. A lot.

In fact, the XO was, to all appearances, a functioning alcoholic who liked to go to the bar on post as often as possible. The problem was, he didn't like to go alone, and he would usually order a number of the junior officers to come along to provide company. As the XO was able to drink a lot of alcohol, these trips for "mandatory fun" could last a long time, and if the other officers matched the XO drink-for-drink (a very risky proposition), not only would they lose the time wasted with him, they would also get home drunk and be worthless to their spouses and family (not to mention a lot poorer for all the money they spent).

Instead of it being a venue for a group of officers to bond overseas, it rapidly became a cause of stress for all the junior officers. They didn't want to go to these events, but when the invitation was phrased in the form of an order, they didn't feel like they had much choice in the matter.

One of the group came up with a solution before things got too far out of hand. He approached the waitress and explained the situation to her. After that, the junior officers got shots of water instead of alcohol, and didn't have to pay for them. The XO still paid full price for all the rounds he bought, and the waitresses kept the extra money. Since the junior officers were drinking water and soda, they could drink them quickly. The XO got drunk quicker, and the junior officers could go home sooner. It was almost a "win" for everyone involved.

But in no way did any of them ever think that it was a bonding exercise, or did they feel more like a team because of it. In fact, the opposite occurred."

It's not teambuilding to make people do things they don't want to, or to be places they really don't want to be. While the junior officers may have stuck together in this case (for their collective defense), this is not the same as being at an event where people are socializing and building relationships. This was an event to be endured, not welcomed.

The XO might have learned how badly the junior officers didn't want to be there if he had ever asked, but he didn't. Effective communications are essential to good teamwork, and being part of a team means listening to the other members. Everyone is important to the success of the group, even if he or she has bad news or something you don't want to hear. A leader would be wise to listen to someone who's trying to help…

Story #19. "Don't Shoot the Messenger"

"I did excellent in the fleet training squadron and had my choice of fleet squadrons to go to when I graduated. I looked forward to squadron life in a "real squadron," as well as getting to fly "real missions" that didn't involve having an instructor breathing down my neck the whole time. I looked forward to my fleet squadron with great anticipation; unfortunately, my excitement waned quickly once I got there. Morale was horrible.

None of the junior officers wanted to be in that squadron, and all of them were actively looking for ways to leave it as quickly as possible. Not only was the CO a man of quick temper and irrational expectations, the XO was worse. He yelled at the smallest imperfection, whether real or imagined. Teamwork was impossible with the two senior officers.

After a couple of months at the squadron, it became apparent the CO and XO knew morale was bad, because they held various events which were supposed to bring us closer together as a team. Unfortunately, the events were just another venue for the command team to yell at the junior officers for another couple of hours, in addition to the ones they already had to endure at work. No one wanted to go to the events, and very few people did, causing the CO and XO to get even more irate.

In an effort to solve things, I went to talk to the CO about the morale problems the squadron was having. I knew most of the issues were based on a perceived unfair distribution of deployment time between the junior officers and the department heads, and I figured the issue could be fairly easily remedied. I also figured that, as the new guy, I was unlikely to be blamed for any of the morale problems; I could bring the issues into the open with the CO so we could all move on.

I was wrong in my thinking, however; the CO actually could, and *did*, blame me. In fact, not only did the CO blame me for the poor morale in the squadron, he also decided I was the ringleader of the junior officers, and therefore I was the one responsible for all the strife in the squadron. For my sin of trying to make life in the squadron better, I was awarded a Letter of Reprimand which went into my official file, greatly reducing my chances for future promotion."

As the saying goes, "No good deed goes unpunished," and in this case, it didn't. Although the officer in question had committed no crime other than wanting a good squadron climate, his promotion potential was vastly reduced, as was his standing in the squadron. Certainly, he no longer felt like he was part of a team, nor did any of his fellow junior officers.

Although this aviator was not punished in public, other people in the squadron were, which can have a tremendously negative effect on a unit's morale…

Story #20. "Lord Voldemort"

"Badgering, belittling, baiting and browbeating (the "4 Bs") were the hallmarks of Army Major "Lord Voldemort," who was named after the villain of the Harry Potter franchise by the troops he terrorized. As the battalion XO, only the commander outranked him, and he loved the power his position provided—nothing pleased him more than to publicly disparage his junior officers and enlisted soldiers to show his power. When he had to deal with his commander and other senior officers, he hid his true nature, presenting a calm and collected professional mien. The times subordinates took their concerns/complaints to the commander, Voldemort always escaped reprimand and continued his oppression—a fifteen-month reign of terror.

As one of his favorite targets, I had the misfortune of frequent contact with the XO when we were tasked with putting together the main operations order for our deployment. Voldemort was supervising, but otherwise not participating. As the XO, he was responsible for directing the staff (all the officers were "staff" for this purpose), and he directed as only he could. We took his "guidance" because we had to, incorporated the non-expletive portions as appropriate and, after many hours of effort, came up with a product to present to the commander. Naturally, the XO wanted to review it first. He was livid. He cursed enough to make a sailor blush, announced to the world our incompetence and, with an amazingly high swear count, gave us explicit directions on how to correct our "abortion" of an operations order. We stayed up late into the night making the necessary corrections.

The next morning we presented the corrected order to the XO. He was livid. Again. This time he was mad because we had changed the order he had liked so much the previous day. We had his own handwriting on some of the originals directing changes, but that was no excuse for changing it. We changed it back far quicker (thank goodness for old copies on the computer), with the knowledge we had wasted hours of work and lost a lot of sleep simply to give him his jollies.

The most telling example of Lord Voldemort's pettiness and spitefulness was his dealings with our personnel officer, who was his favorite target for abuse. During our mobilization, the XO frequently enjoyed what he admitted to one officer was his favorite pastime, publicly berating the Personnel Officer (also known as the S1.) He did this, we decided, because he was a miserable individual who could only feel better when he made others miserable. He certainly succeeded in the latter. He was always after the poor S1. Whether he had actually made a mistake was irrelevant; Lord Voldemort simply enjoyed making the man miserable, and the S1 had no recourse but to take it.

During the mobilization phase, the S1 was one of the hardest working officers we had, and he did a fantastic job (in spite of the abuse). Lord Voldemort's tactics were expanded to include frequent counseling sessions which amounted to little more than belittling bouts. This went on for some months. Finally the S1 broke and refused a "counseling session," and Lord Voldemort pounced. His prey had finally shown weakness, and the XO leapt for the kill. Voldemort charged the S1 with insubordination and refusing to obey a lawful order.

Perhaps Voldemort's reputation was more widely known than anyone expected, however, because this time it was Lord Voldemort who was chastised by the commander of the unit to whom ours was subordinate. The damage was done, though; the S1 was already gone. Another officer, far less qualified, had to pick up the pieces of the personnel shop and do a job he was unqualified for. Although he did a pretty good job, many balls were dropped while he learned his new duties, and the unit suffered as a result."

The deliberate attacks on the Personnel Officer are stark examples of the XO's arrogance and lack of integrity, which led to a complete breakdown of teamwork in the unit. The Personnel Officer's unjust removal resulted in difficulties for all the soldiers in the unit, as the best qualified officer was no longer available to handle personnel issues. There was no teamwork, and the unit suffered because of it.

Chapter Eight
Make Sound and Timely Decisions

One of the strengths of military leaders is that, in most cases, they are not afraid to make decisions. In fact, one of the biggest criticisms of former military leaders in the civilian setting is they are *too* quick to judge. With regard to the "rush to judgment," it is necessary to look at the environment in which they are trained. The crucible of war is unforgiving of leaders who delay or can't make decisions. General George S. Patton noted, "A good plan, violently executed now, is better than a perfect plan next week."

In many cases, it is better to move quickly with the information you have then to wait until you have every bit of information you need. If you don't, the window of opportunity to act may close, precluding *any* action. Because of this, military leaders are trained to rapidly estimate a situation and make a decision, based on the information at hand.

The key to "sound and timely" decisions is evaluating the information you have via a logical and orderly thought process. Some decisions need to be made '*now!*'; however, others will still be timely if they're made two weeks later. When possible, a leader should use the time available to plan for every event that can reasonably be foreseen, rather than jumping to a conclusion on initial estimates. Many times, there will be unintended consequences for hasty decisions...

Story #21. "Why React When You Can Over-react?"

"My first experience with negative leadership happened when I got my initial orders out of the aviation training command. My community had developed a large back-up at the fleet training squadron, and it asked to have no more students sent there until the problem was worked out. As a result, no new aviators were sent for a year. Suddenly, the command realized there were no students available for the next class, and they sent an urgent message to the Bureau of Personnel for a group of students. As a result, 30 of the 35 students in my flight school graduating class got orders to this aircraft (which the majority of us didn't want).

Coming out of the fleet training squadron, we were assigned to our squadrons. The community at this time was in the midst of a transition from dual-pilot to single-pilot crews, and each squadron had a different philosophy on the subject. My squadron's rising CO believed he should have enough pilots to make dual-pilot crews, so a large number of us went there right after they returned from cruise. All of us were assigned to a crew, in all cases dual-piloted crews; the only single-piloted crew was the CO's. As workups for cruise began, it became very difficult to maintain our aircraft carrier qualifications (especially at night) for such a large group of pilots.

Not only did our aircraft have two pilots, we also usually flew double-cycle flights where we would stay airborne for two launch and recovery cycles, so we had half the flights of the pilots in the other squadrons. When you add in the fact that we also had two pilots in each crew, we were only getting one carrier landing for every eight hours of flight time (single-piloted, single cycle aircraft were getting one landing for every *two* hours of flight time.) One result of this was that we started getting scheduled for night touch-and-go's to maintain night currency. One pilot flew the entire flight, performed a touch-and-go, and then climbed to 5000 feet of altitude. Once there, the pilots swapped seats, and the second pilot got the arrested landing.

This was not the optimal way to operate. Things finally got so bad the CO had a meeting with all the pilots and asked if anyone had already made up their mind to leave the Navy, as one pilot was going to be left behind when we went on cruise! One pilot did volunteer; he probably got the best deal of all of us!"

The caveat to "make sound and timely decisions" is that, should you later discover you made the wrong decision, you need to have the courage and mental fortitude to admit your mistake and revise your decision. Trying to stay with a failed decision is far worse than admitting your error and correcting it. And if you're hoping no one will notice you were wrong, you are destined to be wrong again, as they probably already know…

Story #22. "Why Over-react When You Can…Oh, Never Mind"

"Several years ago my daughter went into the Navy as a prospective aviator. The Navy had finally started doing what the Air Force had been doing for years with new pilots. They sent each prospect to a private flying school to get a few hours in light aircraft (usually just enough to get a solo flight). By doing this, many of the students who would have dropped out of the program due to airsickness or other aeronautical inadaptability did so before getting into Primary training, saving the Navy a great deal of money. Unfortunately, one problem came up that no one foresaw. Since fewer students dropped out, each student averaged more flights in primary than planned (dropouts had lowered the average). As a result, a 5-6 month backup of students waiting to start Primary developed. Later, a mishap occurred in Corpus Christi which exacerbated the problem. A T-34C trainer aircraft was lost, and the mishap investigation uncovered a maintenance malpractice throughout the fleet, which resulted in a fleet-wide grounding for several months.

At this point, the Navy's senior leadership decided Something Needed To Be Done. All pre-Primary students were offered the chance to transfer to another community, but not enough volunteered. The next step was to raise passing grades (for Navy Pilots only, Marines and NFO's had no change) for preflight ground school from a historical 80% to 94%, until enough students washed out. This resulted in washing out about 2/3 of each class for a couple of months, at which point the passing grades were lowered back to 80%.

Unsurprisingly, several years later, the Navy realized there were not enough aviators in that year group and tried to get some of the pilots who had washed out to come back."

While making a timely decision is laudable in most cases, a rush to judgment where you don't look at the long-term effects of the decision often ends up poorly, like the two previous anecdotes. In both cases, there was a rush to judgement that resulted in a knee-jerk decision that fixed the current problem; however, it also set the groundwork for a larger problem later. In effect, it only delayed the problem, and when the bill came due later, it had to be paid with interest.

The source of the information often has an effect on the decision-making process. For example, many times a subordinate's opinion will be discarded because, "I know that" or "I know better," whereas we are more willing to listen to the advice of a peer or someone senior to us.

When a senior refuses to listen to your expert advice, sometimes it is more effective to have someone else deliver the news to the decision maker...

Story #23. "Yeah, I'm Not Going to Do That"

"The relationship between the ship's Tactical Action Officer (TAO) and the Flag Watch (the admiral's staff) could best be described as "difficult." A large part of the problem on our ship was the tendency of the Flag Watch to have one solution to every problem: Call the TAO.

One of the cruisers is screwing up the data link picture? Call the TAO.

Can't figure out why the helicopter with the chaplain is late returning from that British destroyer? Call the TAO.

Admiral's shower doesn't have hot water? Yep, call the TAO for that one, too.

During the two months that passed between the infamous "Get Over It" speech and the beginning of the invasion of Iraq, we returned to the Gulf and settled into a routine. That routine ended with the end of the war and, since we had been there the longest, we were released to head for home immediately after the invasion of Baghdad was complete. The timing was perfect; Saddam's statue came down in Firdos Square just hours after we passed through the Strait of Hormuz and into the open ocean.

As we approached the United States, I had the unique experience of clearing "Navy One" into the carrier's airspace. It was one of the rare cases where the President flew aboard the aircraft carrier in a Navy aircraft, making an arrested landing on our flight deck.

As the afternoon wore on, presidential sightings were reported throughout the ship. One of the sailors from my branch reported the President had walked past her in the passageway and given her a hug. My watch officer went back to the treadmill room, only to find a Secret Service agent at the hatch. When he asked when the President might be finished, the agent told him to go on in. The only other person in the space was the President, so they ran together.

I didn't have any direct contact with him during his stay on board, but I did have the opportunity to brief the National Security Advisor and Press Secretary during their tour of the ship. I tried not to take it personally that the National Security Advisor fell asleep during my explanation of the mysteries of Combat. I figured she'd had a long day, and the captain's seat in Combat was awfully comfortable. Or so I'd been told.

In the midst of my discussion, the red light began blinking on the interphone, indicating I was getting a call from the Flag Watch Officer. Excusing myself, I answered the call, and hissed "I'm briefing the National Security Advisor," adding unnecessarily, "of the United States!"

The response was a characteristic sneer. "That's nice. Now, find out why the cruiser is reporting two tracks for contact 4532."

Knowing the cruiser would sort it out within the next 5 or ten minutes on their own, I replied, "Sure," and hung up.

Roughly three minutes later, the phone started flashing again. I glanced up at the display. Sure enough, 4532 was still a double track, and even THIS Flag Watch Officer was not so dense as to miss that I'd neither made a call on the radio to investigate, nor sent them a message on chat.

Without missing a beat in my presentation to the peacefully snoozing National Security Advisor, I reached underneath the table, twisted the cannon plug that powered the phone, and disconnected it.

I'd waited about 269 days of the longest deployment since World War II to do that.

Shortly after being relieved of the watch, we all assembled on the flight deck for a ring side seat to one of the most iconic—and infamous—images of the decade, as the President addressed the crew under a giant banner reading "Mission Accomplished."

Because we were on our way home, many people had already been allowed to leave to go be with their families. I personally had orders to my next command, and was walking off the ship when we arrived in San Diego the next morning. As a result, we had a skeleton crew to stand watch in Combat, and I had to take a second watch, beginning at 2200.

Not long after taking the watch, the phone from the Flag Watch began blinking.

"Combat," I answered (without disconnecting it, this time).

My counterpart during this watch was a much more reasonable guy. "Hey, man; the admiral says you're making too much noise on the flight deck."

I was genuinely stumped. This was a new one.

"What?" I asked, stupidly.

"Yeah, man. He says you're keeping the President awake."

"Uh, are we?"

"I dunno. But the admiral says you gotta quiet it down."

"You know we're breaking down all that stuff from his speech, so the air wing can fly off tomorrow. And also so the President's helicopter can fly off."

"Just see what you can do, okay?"

Sighing, I dialed the phone for the Flight Deck Handler, who was in charge of operations on the flight deck.

"Handler," he answered.

"You're not gonna believe this one," I began. "The Admiral says you've gotta quiet it down on the flight deck."

"Seriously?"

"Yep. Apparently, you're keeping the President awake. Just do me a favor and ask the crew to see what they can do about setting stuff down gently, and not dropping tie-down chains on the deck, okay?"

"They know we're trying to clear the deck for the fly-off tomorrow, right?"

"That's what I told 'em."

"You gotta be shittin' me."

"I shit you not."

"Jesus. I'll see what I can do."

"Thanks, sir."

I should give a little bit of background at this point. The President was being housed in the Commanding Officer's In-Port Cabin. It's a largely ceremonial space, mostly used for exactly the purpose it was being employed now, as a place to bed visitors. The CO almost never used it; he was either in his At-Sea Cabin, just behind the Bridge, or on the Bridge itself. It was an impressive space, containing a precious artifact (a letter) from the President for whom the ship was named.

The In-Port Cabin was in the island, ON the flight deck. The flight deck, being made of steel, is an efficient transmitter of noise, especially when heavy metal objects, such as heavy aircraft tie-down chains, are dumped on its surface. It probably WAS noisy in the CO's In-Port Cabin. However, to all appearances, the President seemed to be hugely enjoying the entire experience of being on board. I rather doubt he was troubled by the noise.

But I digress. Shortly after this little interchange, I could hear the distinctive rumble of a forklift rolling above my head through the steel deck. Sure enough, the phone to Flag Plot started blinking roughly 30 seconds later.

"Combat."

"Flag. The Admiral says no more forklifts."

"Okay... so how are we supposed to clear the flight deck?"

"He says to use hand trucks."

"Hand trucks?"

"Hand trucks."

Awesome. The subsequent conversation with the Handler was considerably less pleasant than the previous one. And, of course, while we were talking another forklift rumbled its way overhead. This necessitated another phone call from the Flag Watch to complain that we were disregarding the admiral's orders.

No sooner had I hung up with the flag watch when there was a rattling, clanking and banging clatter that sounded for all the world as if Marley's Ghost was being dragged across the flight deck.

In the admiral's stateroom, apparently, it sounded like the TAO (me) was deliberately defying him by *increasing* the noise level.

Can you guess what was happening?

It turns out the hand trucks we'd been ordered to use had steel rollers instead of tires. So, by directing us to use them, they'd engineered exactly the outcome they'd been trying to avoid.

In any case, about 20 seconds after the horrendous clatter began, the Flag Watch called.

"Hey, man, the admiral says he wants you to call him right now to explain why you are making more noise on the flight deck."

At this point, I did a quick calculation. It was my last night on the ship. My orders were in hand, my Fitness Report was signed, and I was heading to my last set of orders before I retired. It wasn't impossible for the admiral to do something to make my life miserable, but it was pretty unlikely. And, in any case, ultimately the ship was the Captain's responsibility, not mine.

"Yeah, I'm not going to do that," I informed the Flag Watch. "I'm going to call the Captain instead."

The Flag Watch Officer's response was something between a choke, a sputter and a squawk.

"Don't do that man! He'll have my ass if you don't call him!"

I was sure he was right. I felt bad about that.

About 15 minutes after asking the Captain for help, I got a call from him. "I just finished chatting with the Admiral," he said. "I assured him that my In-Port Cabin was well insulated, and that we'd do everything possible to keep things quiet, but that if we didn't clear the flight deck, the President was going to have a hard time leaving in the morning. He seemed to understand the situation. I don't think you'll have any more calls about it."

And you know—I didn't!"

Chapter Nine
Develop a Sense of
Responsibility among Your
Subordinates

How risk averse are you? With today's "one strike and you're out" mentality, many people are afraid to attempt anything new for fear of failure. We're taught to watch everything we do and say for fear of offending someone, and we're rapidly becoming a culture where people are afraid to try. By developing a sense of responsibility among their subordinates, leaders seek to create a working environment where subordinates aren't afraid to take on new duties and responsibilities.

Subordinates can't grow if they aren't continually challenged, but if the opportunities are presented in an environment where failure leads to punishment, they are never going to want to try. Instead, they will always take the "safe" route, which is almost always the path of missed opportunities. As the saying goes, "Nothing ventured, nothing gained." By being afraid to try, success will often pass them by.

How do you develop an environment where juniors aren't afraid to try? You have to build a relationship where they know it's okay to fail. Start out by giving your subordinates opportunities to attempt duties normally performed by senior personnel and let them know you will accept honest errors without punishment. If they complete the task well, be quick to recognize their accomplishments, especially when they demonstrate initiative and resourcefulness. If they do poorly, correct their errors in judgment and initiative in private and give them the tools and training required to do it correctly the next time. If juniors are immediately punished for everything they do wrong, there is no incentive to try anything new.

One important caveat is that there needs to be a "level playing field," where the standards are the same for everyone...

Story #24. "Board for You, Medal for Me"

"During the early 1980's, a good friend was a junior officer in a squadron at NAS Oceana. During a low-level in the Shenandoah Valley, he hit a power line and nearly tore the vertical stabilizer off the aircraft. He was able to return safely to base, but after an investigation of the incident his CO decided to refer the case to a performance review board, with a recommendation to remove the crew from flight status. The board found the crew not at fault and recommended no restrictions to their flight status.

About a year later, while flying in the Shenandoah Valley, the same CO hit a power line of his own. He did similar damage to his aircraft, and was also able to return to base safely. After this mishap, the CO recommended himself for an Air Medal for saving the aircraft. The medal was refused at the air wing level after the air wing commander reminded the CO of the events of the previous year."

This story is a great example of how *not* to develop a sense of responsibility among your subordinates. Recommending their removal from flight status (i.e., "pulling their wings,") is a career-ending punishment, not one that is likely to make the aircrew try harder. For the leader to then put himself in for an award for doing the same thing only serves to destroy any trust his subordinates might have had in him.

Trust is a key issue in developing subordinates' senses of responsibility. When they know you have faith in them, and that they're trusted, subordinates will give you their all. When it becomes obvious you don't trust them, it is unlikely they will give you their best effort, or care about trying to help you do your job well. Could a missing push pin result in an aircrew missing a flight? Stranger things have happened...

Story #25. "Who Took My Push-Pins?"

"Forward-deployed to a detachment site, military units more often than not benefit from the cohesion of a small team of professionals while they engage in focused operations. The near-absence of paperwork, family problems and social drama at the deployment site usually creates a sense of "one-team, one-fight" that is a tremendous force multiplier.

Our squadron typically sent two or three crews of aviators, along with a maintenance detachment and a small administrative and intelligence support team, to a forward location for several months at a time. The Officer-In-Charge (OIC) of this detachment was a lieutenant commander, a mid-career aviator with 10-15 years of experience. The success of the detachment was dependent on the maintenance team getting aircraft ready to fly and the aircrews successfully executing the mission. The OIC's job was to ensure the smooth flow of information between the two, resulting in mission accomplishment.

Because of the focus on the mission, the preflight preparation spaces for the aircrew had a self-serve "gedunk," which is Navy-speak for a snack bar. The admin support personnel would stock boxes of candy bars, cases of soda, and bags of chips in the gedunk so aircrew didn't waste time wandering down to the local store. There was a simple price list and an unlocked box to pay for the snacks completely on an honor system—like a road-side vegetable cart in rural New Hampshire. The money would come out a little short this time, a little long next time, and over the course of the detachment it would come out about even.

Just about even wasn't good enough for our OIC, though, and he decided to get the accounting squared away. This started with some basic reminders to the aircrew commanders that their aviators needed to make sure they were paying for the snacks they were consuming, *before* they consumed them. Fair enough. But a few days later it became evident to him that his guidance wasn't being followed. He entered the mission prep spaces, marched to the money box, openly expressed his exasperation and immediately demanded the aircrew commander meet him in his office. Once there, the OIC told the commander his crew was "stealing."

Later that week at the barracks, the OIC went to post a notice on the bulletin board near the entrance. He found there were not enough push-pins for him to do so. Again exasperated, he walked to the lounge where the aircrew were eating and watching TV, and he publicly exclaimed someone was stealing his push-pins. Being good naval aviators, the crew accepted this challenge and, before they retired for the evening, removed every push-pin from the building.

These two seemingly minor incidents worked to form distinctly negative impressions on the aircrew. First, the OIC was not focused on the mission; instead, the OIC's energies were focused on tangential and largely irrelevant issues. Worse, the OIC did not trust the members of his team. He openly accused his officers of dishonesty. Worse than not being a team player, he became an outsider who wasn't even considered to be *part* of the team.

The frustration generated by these minor incidents led to a downward spiral of events between the OIC and the aircrews. The aircrews felt the OIC wasn't focused on the mission, while the OIC became increasingly frustrated his aircrews "didn't get it." All of this needless drama detracted from the mission. Every moment spent complaining about the OIC, and every meeting the OIC scheduled to resolve problems like a minimal amount of missing gedunk money, was time not spent focused on executing the mission.

The culmination of these events was the encounter between an aircrew commander and the OIC at the end of the month. According to naval aviation rules, there are time limits imposed on aviators for safety—limits on the number of hours flown in a 24-hour period, limits on crew rest before flights, limits on alcohol consumption and limits on the number of hours an aircrew could fly in a month. What should have been a coordinated team effort to monitor these trends to minimize their impact on mission accomplishment turned into a Mexican stand-off. Ultimately, the OIC and aircrew commander reached the inevitable position where the OIC needed a mission flown, and the commander informed him the crew was beyond limits and could not do it. The OIC was beside himself. In his mind, the aircrew was out to undermine all that was good in him, the Navy and the world, and he crumpled up the flight schedule and threw it at the aircrew commander before storming out the door. The mission was canceled."

This might have been an important flight, but we'll never know; it was lost due to an unnecessary conflict. Better teamwork would have helped here; had the OIC and the aircrew commander been working together ahead of time, a solution could have been found and implemented; the flight *could* have been completed. Instead, the OIC's lack of trust in the aircrew commander destroyed the sense of responsibility he had, and caused the crew to miss the mission. Could something as small as a few push pins cause a crew to miss an important mission? Yes.

Is it hard to develop a sense of responsibility in your subordinates? In most cases, it isn't difficult at all as most people want to do a job well and help their organization succeed. Usually, all that's needed is a little trust…

Story #26. "What Part of "I'm Doing You a Favor" Do You Not Understand?"

"I spent five years at an analysis organization affiliated with a military education institution. The distance education program of this institution was always looking for people from the rest of the organization to go to their remote sites and teach a class or two so students would get more than a single instructor for eight months. The instructors who put together the courses, not the local instructors, were responsible for finding the guest instructors. I was on the research side, not the teaching side, but I took being asked to teach a class as a compliment.

Now, you would think the distance education program would do everything they could to make this an experience people would want to repeat, and indeed, most of the people in that program were a pleasure to work with. The one exception was the deputy, who I will call Ford. I later discovered Ford was the single reason why they had a hard time getting volunteers—NO ONE liked dealing with Ford.

My first experience with Ford came after I had already signed up for the program and was starting to make travel arrangements. I had originally said I would be happy to go to Norfolk and teach three classes, but only if I could combine it with a trip I had to make to Washington, DC, anyway. No problem, I was told; just talk to Jim (the departmental secretary) about the details. I asked Jim what the regulations were about driving my car, driving a rental car, and taking the train, and we discussed the options (flying didn't make sense, as I was going to two places only 150 miles apart, and would need to rent a car anyway.) I eventually settled on driving. At the recommendation of colleagues, I decided to rent a car.

A couple of days later, I was talking with the guy who recruited me for this duty and told him I would drive down to Norfolk from DC the morning of the event. "Oh," he said, "our policy is for you to drive down a day early so you have time to recover from a transportation emergency enroute." Good to know, I said, and immediately fired off an email to the secretary to change my hotel reservation to come in the day before.

"This is the LAST CHANGE allowed," was the response I received from Ford. The rest of the response was pretty hostile, too, keeping in mind I'd never actually met Ford nor exchanged emails with him. I responded that I was surprised by the tone of the response, given this was the only change I had requested.

"No," I was told, "you've been constantly changing all of your travel arrangements, and you have made unreasonable demands about driving and side-trips to DC and the like."

I was befuddled as I'd made no such changes; the "side trip" was clearly billed up front as the quid pro quo for my participation, and besides, their own instruction sheet said "we expect that you will take advantage of this opportunity to visit other places in the area." The only change I had made was the hotel, and that was to bring my plans into harmony with *their* policy, which was not in the materials they had initially given me.

I replied to Ford I was confused, and why that was, at which point he told me that my characterization of what was going on was wrong, and that I had been constantly making changes. Not "that's not what I understood," but "you're wrong," which I thought was pretty bold coming from someone who until this point was not part of the discussion. I considered pulling out then and there but, knowing how hard it was for my colleague to get volunteers (and now I know why!), I gritted my teeth and said, "Roger, no more changes."

I went on the trip. While I was in DC, I met some other people who were also going to do the "visiting professor" thing, and we discussed how we were going to approach the problem. The lectures went great, I got my DC meetings done and came home and filled out the expense report. Glowing reports came back from the instructors I had helped out in Norfolk.

Fast forward a year. My colleague asks me to do this AGAIN (remember those glowing reports from last time?) and, knowing what a hard time he had finding suckers to do this, I plastered a smile on my face and said, "Sure! Anything to help a colleague." The next day, I got a call from the colleague, asking me to come over to his office.

I thought I knew what was coming, and I was right: Ford had banned me from ever being a visiting instructor in the program again. "I understand," I said, both angry and exuberant at the same time. I *think* I succeeded in keeping the smile off my face...but I never asked, so I don't know.

Epilogue: You may recall Ford was the deputy department chair. Shortly thereafter, the term of the department chair was up, and a new chair needed to be selected. Ford applied for the job, but didn't get it, and there were reports indicating the administration understood that if Ford had been hired, they would have had to replace the entire department. Sometimes, bad leadership is its own reward."

By banning this person from participating in the program, the command lost out on a valuable asset and set the conditions where no one else would want to participate in the program either. Not only does he make it difficult for juniors to seek out new responsibilities, the deputy in this story doesn't seem to understand the unit's mission and how the visiting professor program augmented it. Understanding the unit's mission, capabilities and its proper employment is crucial to its success, as we will see next.

Chapter Ten
Employ Your Team or Organization in Accordance with its Capabilities

One of the biggest blows to morale occurs when a unit is employed in a manner that isn't in keeping with its mission or its capabilities. No one wants to be misused or used badly, and no one wants to be put in a position to fail, but that is what often happens when a team is given a task they don't have the tools or training to accomplish. If you don't know how to complete a task or you don't have the right tools for it, you are unlikely to perform it well, especially the first time.

I've got firsthand knowledge of this. The last time I called my detailer, he told me he had orders for me to go be a convoy commander. My only choice was whether I wanted to do it in Iraq or Afghanistan. As the year was 2007, neither of those were particularly pleasant choices...for the people who did that as part of their normal jobs, much less for an aviator. I didn't have the training to do the job well, and I didn't want anyone to get killed because I wasn't prepared for the job. I decided to retire from the Navy, rather than take those orders.

When I asked why we were filling these positions, the detailer told me the Navy had taken over the convoy commander positions because the Navy "wasn't getting enough people killed." The other services had lost a number of people in these countries, especially doing convoy command duties, so they had more credibility with congress when those services went to ask for additional funding. The only reason the Navy was sending people to these jobs was so that we could get some folks killed, too, so we'd have more "cred" with congress.

Several thoughts immediately went through my mind. 1. WTF??! 2. If true, I didn't need to know that. 3. That's probably *not* the best marketing tool to use with prospective candidates if you want them to take the job. 4. Thank God I'm no longer in an organization that thinks things like that! How do you show someone what little value they have to management? Tell them they are more valuable dead than alive and doing their daily job. 5. And one last, WTF??!

That was absolutely the *worst* thing you could say to someone. Even *if* the organization's management is so cold-blooded they think things like that, there's *no way* you should say so if you <u>ever</u> want your juniors to have faith in the organization again. My faith in senior management went straight to zero, and I was happy to retire rather than take the orders.

Why would I want to stay in an organization that was actively trying to kill me? How many others got out of the Navy once this became known? It's impossible for me to know, but I am aware of several others who did. This decision cost the Navy many of its "best and brightest," as those who could leave when this became known, did choose to leave. Send me over to fly in a dangerous area? Fine, that's what I signed up for. I would have been happy to do it. I was trained and had the tools to do it. But to go do something I was ill-prepared for, just because the Navy needed to get some people killed? No thanks.

Sometimes, there is no other choice than to employ an organization outside its normal purpose and capabilities; the unit may be the only one available to do the job, or the only group that has even some of the requisite capabilities. In that case, it's up to the leader to explain why the deviation from expectations is required. Before this can happen, though, leaders need to know their own units' capabilities, so they can assign reasonable tasks to their subordinates...

Story #27. "They Should Do What We Tell Them"

"I met the captain after he had been passed over for admiral and put out to pasture at the Reserve Officers' Training Corps (ROTC) unit at the college where I was stationed. It didn't take long for me to see there were going to be problems. He lost his temper so often every officer actively avoided him. He would ask the junior officers to go golfing with him in the morning, and then after lunch he would yell at us for being lazy and slacking off on the golf course! He also didn't work well with civilians; he called them names behind their backs and constantly talked about how overpaid they were. He absolutely refused to work with them on their own terms.

The problem started when I took over as "Nurse Coordinator." The Navy had just started taking nurses into ROTC and required them to fulfill all the ROTC requirements. Unfortunately, the Navy didn't look to see if a student nurse's schedule and the ROTC requirements for midshipmen could be completed at the same time. They couldn't, at least not at my university.

Seeing a problem, I called my nurses in, pulled out the checklist and asked them to plan out their classes for the next two years. It turned out they could not take the military history course due to their rotations in the hospital, nor would they be able to schedule any of their senior-year ROTC classes.

Like many universities, we had a history department class that students had to take to fulfill ROTC requirements. The Nurse candidates could have taken it the year before, but now it was too late. I called the Navy's nursing education department in Pensacola and asked if anyone had thought about history and had a plan to get these midshipmen commissioned. They hadn't. Worse, the course had to be completed for them to earn their commissions.

I took it to my captain, along with a solution I had devised. Instead of listening, he screamed at me. His face got so red I thought he might have a heart attack. I wasn't sure what had happened; for some reason, I was in trouble for finding a Navy-wide problem and having the audacity to propose a working solution.

He ordered me to get the history department to arrange a course for the Nurse candidates. When I explained that his order violated the university's procedures, he yelled at me even more and ultimately gave me a letter of reprimand for not following a direct order.

I did, however, follow the order and make the call. I hated every minute of it. And though I was not laughed at by the history department, I was given a thorough "lesson" in how university departments work with each other. It could have been uglier, but they let me down easily. The dean of my department, however, was not at all happy, and he wanted to know why I was making requests that did not go through him.

The lesson learned was that our small command was the guest of another organization, and a civilian one, at that. You can't order an outside organization to do something if you have no authority over them. Although my captain complained, "we pay them, so they should do what we tell them," in truth, the only thing we could have reasonably done was ask them for their help."

The key to employing your team or organization in accordance with its capabilities starts with actually knowing what those capabilities are and how to use the team. Unless you've been with the unit a long time, though, or worked your way up through it (i.e., "came up through the ranks"), you may not be familiar with all of a unit's capabilities or how to effectively employ your people. That's okay; there's a way around this.

Ask.

When engaging in a new project, talk to the folks who work for you and bring them in on the planning. They know best what they can and can't do, as well as the tools and training they need to accomplish the new mission. Put that knowledge to work for you by actively seeking it out. Don't want to ask? You could just go it alone...

Story #28. "Emperor Happy"

"This story is about an Air Force major who we will call "Happy," who was the XO of the command. A morose, humorless individual, he was also something of a tyrant. He was an empire builder, always striving to ensure his position was secure and things were done his way, regardless of whether or not it was in the best interests of the unit. His penchant for browbeating and creating a hostile work environment led him to be dressed down by the commander on a number of occasions. This had an ameliorating effect...until the commanding officer was promoted out of the unit, and Happy took over; at that point he reverted to his old ways. The result was easy to predict: plummeting morale, airmen looking for transfers out of the unit, and personnel who tried to maintain as much distance as possible from Happy, even at the expense of career advancement.

During the time I was there, Happy's unit was part of a joint task force with the Army, which had been housed in temporary quarters for several years while waiting for its permanent facility to be completed. Unfortunately, Happy was instrumental in organizing and planning the new structure, and he made sure the building took the shape *he* wanted it to have, not one that would have best served the tasks to be undertaken within its halls.

None of the functional analysis that should have been conducted ever was; instead, Happy chose where personnel and capabilities were to be placed. His aspirations to create a legacy instead of a truly functional structure left the unit with a building that could only be called "inefficient," at best. While the unit was able to make it work, it was not conducive to the type of operations they ran, and staffing was problematic due to a lack of meeting rooms, training rooms and offices. Many of the ancillary functions had to remain in the old, "temporary" facilities due to the lack of space. There was plenty of space available in the building; it just wasn't properly used."

To the men and women of the unit, the new building was an inconvenience and an annoyance. Every day when they came to work they were reminded of the fact the building could have been so much more...but wasn't because the leader didn't care enough to make its form fit the functions of the unit. When the leader's same lack of caring impacts family life and off duty time, it becomes more than just an annoyance; it becomes an enormous morale buster...

Story #29. "I'll See You on Monday"

""'I want maintenance to work this weekend," said the CO. I was the maintenance officer of a squadron that had just come back from the latest war. The men and women of the maintenance department had done an outstanding job on deployment, with only one flight missed in three months, and they were looking forward to some time with their families.

I was unaware of any unmet commitments, so I asked, "Okay, Skipper. What are we supposed to be working on?" All of our aircraft were in an "up" operational status, except for a couple that were undergoing routine planned maintenance, and there was nothing on the schedule that required them to be "up." The planned maintenance was scheduled to be completed on Monday of the next week, and there weren't any operational commitments for several weeks. I thought I must have missed something on the schedule.

"Don't get smart with me," he replied. "I told you maintenance is working this weekend; now make it happen!"

"Yes, sir, I will," I replied, "but it would help if I knew what we were supposed to be working on. Is there something I missed or somewhere we're not meeting our commitments?"

"Darn it, I told you to quit being smart!" he replied (edited for profanity). "Maintenance is working this weekend. Just shut up and freaking do it."

At that point, I realized I wasn't going to get any actual direction or guidance; whatever was driving the requirement to work on the weekend was going to remain unknown.

I went down to Maintenance Control and gave the Master Chief the happy news. After he ran out of swear words, he finally asked, "So, what the heck are we supposed to be working on this weekend that's so darn important?"

"I don't know," I replied, "so we're going to work on the things we can. Since we have to be here, let's get the two aircraft finished up and work on people's qualifications." There had been less time during the war to train and get people qualified for various tasks; the focus had been on preparing aircraft for combat.

So, turning lemons into lemonade, we all came in and did what we could for the weekend, even though no one's heart was really in it. Planes got fixed, training was (somewhat half-heartedly) held and then the weekend ended and we went back to work for real. The CO wasn't seen at the squadron all weekend, nor did he call or indicate in any way he cared about what we were doing.

What he did care about, I found out later, was walking into the air wing commander's office that Monday morning, where he proclaimed himself to be the only CO in the air wing who had "all of his aircraft in an "up" status," which he made sure the air wing commander and the other squadron commanders were well aware of. The working weekend had been called for one purpose only, so he could claim superiority over his counterparts."

The squadron wasn't employed within its normal capabilities or in the manner it normally should be; it was employed for the sole purpose of self-aggrandizement and helping the CO get promoted, at the cost of its members' morale and quality of life. My takeaway was that, if you can't explain the tasking to your subordinate, it probably isn't something you should be ordering them to do.

Chapter Eleven
Seek Responsibility and Take Responsibility for Your Actions

Responsibility. The military is built on accountability and taking responsibility for your actions. Unlike civilian organizations where a leader can get out of the blame for something that happened by saying, "I didn't know that was happening in my organization," the commanding officer of a military unit is ultimately responsible for everything that goes on in the unit, regardless of whether he or she knew about it. It is up to the military leader to set the right standards and exercise the appropriate leadership to ensure the unit's climate is such that "bad things" don't happen on their watch. The leader (regardless of rank) is responsible not only for his or her own actions, but also everything the team does or fails to do. That's the way it's supposed to work, anyway...

Story #30. "The Five-Year-Old Aviator"

"I was the Navigation Officer in the Operations Department of an A-6 squadron. It was a great job for my second assignment as a junior officer, and I loved it.

Unlike most other officers, though, I had never wanted to be a career officer, and I was on my way out of the Navy. I wanted to go through ROTC, become an officer and serve a few years, and then head back to being a civilian before age 30. This career path was frowned upon in the military; even saying it got me in trouble. I probably should have played the game better, but I didn't have it in me to be something I wasn't.

After Desert Storm, every other officer in Operations rotated out, and I was left behind. I was the senior lieutenant (LT) in the command; all the new personnel were junior to me, including the new Assistant Operations Officer. At the time he was the "golden boy," and being assigned a lieutenant commander's job as a relatively junior LT went to his head. Instead of treating me as an experienced team member and a resource, he treated me as a pariah. He would not listen to any of my suggestions, and he certainly didn't ask for help. To him, the fact that he was 'promoted' over me obviously meant he was superior to me and didn't need any help...especially mine.

To make matters worse, the new Schedules Writer and I had an acrimonious relationship. His desk was next to mine, and he openly mocked me. He had taken the place of an exceptionally intelligent junior officer. At the time, I was also the back-up Schedules Writer, and I had been writing schedules for more than a year. That experience was scoffed at by the new Schedules Writer; I was 'just' the Navigation Officer, and he was 'hand-picked' over me for the more high-profile job of Schedules Writer. What could these two rising stars ever learn from me?

The squadron was assigned a particularly tricky practice mission, which involved flying through and around restricted areas and coordinating with the Army. At that time, joint service training was still in its infancy, and there were *a lot* of issues every exercise had to work through.

Now I was admittedly surly, and I wasn't very happy with the new situation. With all of that said, I was still a U.S. Navy lieutenant on active duty in a combat squadron; I did my job, and I did it well. If something was asked of me, it was completed quickly and professionally.

I had planned and filed the paperwork for a number of training missions over the previous year, and I had experience with what needed to be done. The previous Schedules Writer and I had worked as a team, even if I was mostly his assistant. By the time this exercise rolled around, though, I was the only one left in the department who had ever been part of a joint exercise. I could see they were all in way over their heads, so I offered my help to the Assistant Ops, but he told me to shut up and go away. Literally.

The Assistant Ops Officer jumped at the chance to show his superior skills in all things Naval Aviation, and the Schedules Writer smirked the entire time. They took the mission on as their sole project, and the Operations Officer agreed to let the two 'stars' have it.

Well, you know the ending. It went badly. Really, really badly. They didn't file the correct paperwork with the Army or the Federal Aviation Administration. The two-plane flight flew through restricted air space, and a team of very senior officers were calling both the squadron and our air wing commander to discuss it *while the crews were still in flight.*

Things roll downhill, as it's said, and the 'discussion' the CO had with the Operations Officer was incredible. It could be heard a long way off…even from behind two closed doors.

When it was over, the Operations Officer came back in and called for a closed door officers-only meeting. He took full responsibility for the fiasco and apologized to me. He then turned to the Assistant Ops and said, "I gave you the chance to be all that is said of you. I gave you the chance to excel and didn't micromanage you, but you didn't use your team; I watched you be a jerk to the Navigation Officer when he tried to tell you how this could go wrong."

He went on to say that, until further notice, everything in the department had to go through me, as I was the most experienced officer in Operations. I would help get them pointed in the right direction. He even went so far as to make me sit at his desk, while he moved to a chair on the side of the desk.

The Assistant Ops could have admitted fault, and wrote it off as a rookie mistake and moved on, but he didn't. Instead, he became even more of a jerk. He absolutely refused to talk to me or even acknowledge I was sitting in the desk across from him. He wouldn't even hand his paperwork to me; he would give it to the Operations Officer who invariably handed it to me for review. If the Operations Officer was out, he either waited for the Operations Officer to return or just threw it on the desk condescendingly."

Not everyone wants to stay in an organization or a particular position, but that doesn't mean they are disloyal. There are even times when people are happy in a position and aren't looking for promotion. It's important to remember experience comes in many shapes and sizes, even if the people involved aren't viewed as "stars," or if they are junior to you. Refusing to listen to an "old salt" can be hazardous.

Being a good leader requires the ability to listen to people you may dislike. They may even be people who are, in general, not as talented as you are, but have knowledge of a specific task or area. Sometimes they may be very talented but surly. Filtering what they say to you and listening to any advice they have to give may be the difference between mission success and failure. Listening to all voices IS the job of the decision maker.

Two great questions to ask all of your subordinates, which also help integrate them into the team, are, "What do you think could possibly go wrong? How would you plan to avoid it?

All of this assumes, of course, that you actually *care* about what you're doing...

Story #31. "Disengaged"

"My father was a Marine's Marine. If you asked him what his perfect vacation would be, he would respond, "to storm a beach facing overwhelming odds." When he learned of my plans to enter the Navy rather than his beloved Marine Corps, he emphatically stated, "I'd rather have a daughter in a whore house than a son in the Navy!"

That's the backdrop to a moment in his career when he was sent to his "dream orders." Following 13 months in Vietnam, in which he led a company that sustained high casualties in the Tet offensive, he was going to be the XO of a Marine embassy detachment oversees.

Upon reporting, he learned his CO was a full blown combat-avoider, who looked to my father like "10 pounds of crap in a 5 pound bag." Unit morale was bad; there was general dysfunction and poor leadership on all levels, including the unit's gunnery sergeant, who was known to be an alcoholic. He routinely partied with the men, who were also fresh from tours in Vietnam.

My father's dream evaporated; he was attached to a unit that was everything he *didn't* want. As is too often the case in the military, he also took over his position without turnover from his predecessor. He had no direction from his chain of command and no training on how to run a security detachment in a foreign country. Before doing anything, he inspected and evaluated his men only to find a total lack of military bearing and discipline. Infractions ranged from drug use to reporting for duty while under the influence of alcohol. An even greater concern was the evidence he found of major security infractions as his troops routinely allowed access to the barracks to any number of foreign 'ladies.'

But he was a Marine, so he took charge. He categorized all the unit's issues in his first 30 days, and, rather than present his CO with just a list of problems, he created a plan to fix them.

He made an appointment with the CO, who did not have an open door policy, excited to give him the brief which would get the detachment back on track. Upon entering, the CO did not greet him, nor did he even look up from his paperwork. My father later learned it was probably a crossword puzzle, but his view was obscured by the mounds of unfinished correspondence. As my father progressed through the brief, he was not sure if his boss was even listening as all the communications went one way. At the point he hit the unit's biggest problems, the CO looked up for the first time. He seemed somewhat surprised to see my father and interrupted him with the questions, "What are your thoughts on the upcoming costume party? Do you think I should go as a knight or a bunny?"

Dumbfounded, my father waited for the punch line...only to see his boss retreat back into his paper work. My father was pushed past the breaking point and, without thought of career or retribution, quickly responded, "Why don't you wear your dress blues and go as a United States Marine?" With that, the meeting ended, and my father left the major with an expression of amused surprise.

The meeting was liberating for my father as he realized where all the unit's problems were coming from. His CO had abdicated responsibility for all the unit's duties and responsibilities. In the absence of orders from his CO, my father took the initiative to perform the actions he believed a competent leader would have, had there been one present. He set about executing the changes he had proposed in his report, knowing it was easier to seek forgiveness after the fact than it would have been to get permission in the first place.

Although worried his fitness reports would suffer, my father found that wasn't the case; the same character flaws that led his CO to be completely disengaged also led him to fear writing anything negative. When it was time for his next set of orders, the CO had moved on, and my father finally got his dream orders, this time to train young lieutenants for combat at The Basic School, which confirms the old adage "no amount of perseverance and hard work can overcome the power of good luck and timing." "

While it worked out this time, on most occasions being this disengaged *won't* have a liberating effect on your subordinates; instead, they will usually take a lack of caring on the part of the leader as an indication they should also not care. If you can't care about defending an overseas position, I'm not sure what you *can* care about.

As was mentioned earlier, one of the keys to seeking and taking responsibility as a leader is in setting the right climate wherein everyone knows what is acceptable. Many times leaders give tacit approval to actions that aren't permissible, whether or not they would have actually approved of the actions. When leaders set this kind of climate, excesses can occur...

Story #32. "Crossing the Line, While Crossing the Line"

"Back in the early 1990s, I was assigned to a small surface ship that conducted a Western Pacific and Middle East deployment from our home port in Pearl Harbor. In the final month of this deployment there was a "Crossing the Line" ceremony that resulted in a medical evacuation. Things happen sometimes, and nobody wants to see a shipmate get hurt, but the lessons in this case go beyond a safety procedure being implemented for the next event. This story is a case study in leadership, and how the actions of leaders can have wide-reaching unintended consequences.

In the current era, or at least in the last decade, a sailor deploying on a surface vessel could expect to spend 6-10 months deployed, almost all of it "on station" in whatever region they were deploying to support, with a total of 4-6 port visits during the deployment. My last two deployments were on an aircraft carrier in 2009 and 2010 to the Middle East. Both together totaled less than two weeks in port during the six and a half, and then eight months, we were deployed. In contrast, my first deployment was scheduled for 14 separate port calls; we spent more than two of our six months deployed in foreign ports.

This was my first ship and my first deployment, which amplified the excitement of an already new and amazing experience. We had come out of the shipyard several months before; during that six-month yard period a number of major refits were completed, as well as a good amount of extensive and long-delayed maintenance. While this was all good and necessary, within weeks of our entering the dry dock Saddam Hussein decided to roll across the border into Kuwait, and we missed Operation Desert Storm in its entirety. Although there were still a few skirmishes on small islands in the Arabian Gulf when we finally got there, six months after the cease fire the real war had ended.

The deployment in question went like most, quickly in hindsight, but achingly slow in real time. We spent weeks conducting independent operations in the Arabian Gulf at its northernmost point. We logged endless amounts of time watching rigs burn, patrolling our sector, sinking crippled vessels and barges, and sometimes even fishing bodies out of the water for internment.

I learned an incredible amount about the Navy, seamanship and leadership in these months, but the biggest lesson was yet to come when we went for a quick down-and-back to Bali, Indonesia and Darwin, Australia. Everyone was excited, not only because of the ports, but also because we would finally have a "Wog Day." When ships cross the equator, or "The Line," there is a long standing tradition of hazing the "Wogs" (the sailors who have never crossed the line before) by the "Shellbacks" (those who *have* crossed the line). The "ceremony" involves physical exertions, corporal punishment and a lot of silly behavior (leg shaving, garbage crawling and cross-dressing at a minimum). At the time, we probably felt the be-havior was more appropriate, or at least less inappropriate than we would now, as warships were single gender. Now I wonder, "Who the heck thought that up?" Despite the weirdness, Wog Day was intended as a rite of passage to help the crew bond. While meant to be a little scary to the Wogs, it was also supposed to be a good time. As one would expect, it required a good bit of supervision and ma-turity to pull off well.

Which brings us to the Wog Day preparations. The day before Wog Day, we had a Captain's Call. Information was passed and awards were presented for noteworthy accomplishments. At the end of the ceremony, the "Wogs" were dismissed and a quick brief was given for the following day's events. Unfortunately, this is when it started to go wrong. At the close of the brief, the Commanding Of-ficer made an offhand comment in a pep rally-type voice to the ef-fect of, "We're really going to show them tomorrow!" I am a hundred percent certain his intent was to get the guys excited about a fun, "let's blow off some steam," event, but the train was coming off the rails.

Wog Day dawned and the excitement grew as the day progressed. There was the expected razzing of the new guys by the "old salts," and the even more foreboding silent looks by some of the older guys. Some Wogs walked around visibly nervous, some thumped their chests and others were just along for the ride. By the end of evening chow, the air was electric. The official event was set to begin after midnight, but by early evening the first-term sailors were asked to leave the spaces and make themselves busy elsewhere. In the in-tervening time, these additional spaces were made ready for pregame trials and punishments. These were not "officially" part of the line-crossing ceremony, which itself consisted of several hours of hazing as one crawled the length of the ship.

I was on watch that evening in the Combat Information Center (CIC). We were in a lightly-trafficked portion of the Java Sea. While it was very busy on a worldwide scale, traffic was negligible as compared with crossing through the Straits of Malacca, which we had done the week before. During the watch, the tension was apparent, but certainly not so thick as to impact watch-standing ability. As the watch section we relieved departed, they were all instructed to get chow and then report to the "pregame spaces" I mentioned earlier. As it turns out one of these spaces was at the bottom of a ladder well four decks below CIC. As people transited through our space we could hear the yelling from four decks below, both the Shellbacks and the Wogs, and one set of voices was clearly happier about it than the other.

About halfway through the watch, the CO walked through CIC. He checked on our view of the navigation picture and, unable to resist the temptation, asked if everyone was up for the "festivities" to come. I have thought about the next several moments many times over the intervening years. While I have also reflected on what transpired over the hours that followed as well, this is the point where my own actions entered the narrative, and frankly I'm not proud of how it started.

The CO's question to the junior sailors was tinged with just enough flippancy to make it clear to me he was enjoying their discomfort. I let my mouth run away, a not uncommon occurrence, and gave a snarky-enough response that it was clear to him I didn't appreciate the cat-and-mouse game, but not so snarky as to be outright disrespectful. I mention the interaction for several reasons. First, I realize now I was wrong on multiple levels. Most importantly, who was I to decide the Captain was "too flippant" and then take him to task, no matter how subtle I thought I was being in front of our crew? At a pragmatic level, I had also just challenged his authority, making me a target. Second, after giving me a two-second glare, the CO opened the hatch, letting in the screams from four decks below, turned and smiled at me and said, "Maybe you won't be so smart when you get off watch." It was the exact response I would have given, and it became relevant to future events.

When the watch ended, I went to my bed, but before I could close my eyes I was summoned by other Wogs who had completed their turn in the furnace. Apparently, the fee for leaving was sending the next victim to fill your place. My next hour involved a lot of yelling and being yelled at, and a prolonged interaction with several shipmates' shillelaghs. For the uninitiated, in the context of Wog Day, a shillelagh is a three to four-foot length section of three to four-inch wide, canvas-sheathed fire hose. It is swung like a club in the shorter version and baseball bat in the longer version. They are used to essentially spank Wogs as they crawl the predesignated course running the length of the ship. There's some technique involved, as a swing to contact has a slapping effect that creates welts and noise. A swing through contact, especially to the rear end, will launch the victim forward and create deep bruising and striations. I suspect initially they were part of the general taunting and harassment Wogs received while crawling along; however, in our case, we spent close to an hour during this "pregame" with a dozen or more guys armed with them in an enclosed space, receiving both types of blows.

Ironically, I look back on that night somewhat fondly. Personally, I never felt like I was in danger. Don't take me wrong, it hurt, and I walked funny for a week due to the bruising, but I trusted my shipmates and had a relationship with them that made me confident that, while I "was going to get mine" (which they had in fact promised for days, if not weeks prior, and the CO had just implied, as well), I knew I wasn't going to be seriously injured. In the end, that is exactly what happened. While everyone took their turns, I continued to let my mouth run away. When three of the bigger Shellbacks, who had "tricked out" shillelaghs (with diamond-shaped holes cut in the end that amazingly left diamond-shaped welts), asked me what I thought of my beating, I asked them in reply if "that was all they had." They went at it again, and then broke to ask if I'd had enough. I replied, "You should send your husbands next time." The next stroke launched me about eight feet forward, headfirst into a bulkhead. When they asked, "How about now!?" I mumbled a woozy and humbled, "I'm good," which got big laughs, and they sent me on my way. Days later, the sailor that landed the big blow pulled me aside and told me that my bravado was pretty funny, but he was worried it was drawing a crowd. If I hadn't kowtowed at that point, the next one probably would have really hurt me.

I departed for my room, and just prior to dawn all the Wogs were mustered on the bow of the ship to commence the ceremony. There was firefighting foam, silly costumes and more screaming. There was also some beating, but nothing like the night prior. As we departed the forward area of the ship, the Captain came over the public address system and announced the ceremony was being suspended because we needed to launch our helicopter on a medical evacuation mission. Our initial presumption was a civilian in the area had been injured on another vessel, and we were launching the helo to pick them up and take them to medical care, a not uncommon occurrence. While we were concerned for the victim, we were not thrilled the word the CO had used was "suspended," implying the Shellbacks could rest up while our adrenaline wore off, and we would start all over.

Following the launch of the medevac mission, the CO made a second announcement clearing the issue up. One of our junior shipmates, a guy less than six months out of his initial training, had been injured and had to be sent to the hospital. He had passed blood with his urine just prior to the ceremony or had to stop during it to do so. The fear was a shillelagh strike to the kidney had caused the damage, which is possible as he apparently received a good bit of special attention the night before. However, other reports from sailors in the division indicated he had hurt himself in a fall during heavy seas several days prior. In retrospect, that sounds ridiculous and cliché. But I trusted, and still do trust, the witnesses who also didn't deny the Wog events could have easily made it worse. The Captain then announced Wog Day festivities were cancelled. The story, however, doesn't end there.

When the ship reported the injury and medevac (as was required), the news raised a good bit of attention. In the current era, and especially in the last decade, the Navy has made an enormous effort to end hazing and root out incidences of sailors hurting other sailors. Even at that time, nobody wanted anyone to get hurt, but it would likely not have received the high-level attention within the hour as it does today. Our case was slightly different, though, as it was not our first incident. During our deployment, we had experienced several others. We had an attempted suicide, which followed multiple open statements of intent that were not acted upon by shipmates, we had turned on a radar with men aloft and we had fired our close-in weapons system without receiving permission to fire first. All of these were bad, but this was the first after transiting into a new area of operations. The new fleet commander recognized the trend and acted immediately, demanding the Captain contact him to give a full report.

The next morning, as we approached the anchorage off Bali, we were directed to man the rails in our summer whites. This was unusual, as we were far enough out our uniforms would be virtually invisible to anyone on the beach. But, looking forward to a 2-3 day visit made it worthwhile. Even more perplexing, we were directed to man up an hour early, so we had a good deal of time standing there to soak in the view. Once anchored, we were left at the rails for another twenty minutes while First Division, who conducts the actual anchoring, was called to prepare the Captain's Gig (a small boat we carry with us) for a run to the beach. After the boat was prepared and lowered to the water the Captain came on the public address system and informed us he had been summoned to the beach to call his boss and explain why "we" had performed so poorly. He was cancelling all liberty so we didn't cause any more problems as a punishment, and we were then moving on to our next port visit in Australia. The crew, on the whole, was neither amused nor impressed.

When we entered the harbor in Australia later that week, there was a large crowd on the pier. It was a small town and, relatively speaking, we were somewhat of a big deal. Unbeknownst to us, the local media had arrived to cover the story several hours early, and to pass the time they had all fallen upon the only member of the crew available, the sailor who had been medically evacuated and was now returning to the ship. As one might imagine, having been recently injured, and having a personality that led his shipmates to take their ire out on him in the first place, he was the worst possible candidate to portray the ship in a positive light as we steamed into port. The next day's headline didn't announce our triumphant visit, but instead proclaimed in two to three-inch bold letters, "Sailors Receive Brutal Bashing." As receiving any negative press *also* requires a report to the fleet commander, and this one mere days after he had summoned the boss to explain the last event, the fleet commander reached the end of his rope with our ship. Within days, a team of investigators and lawyers arrived to get to the bottom of our problems, which is when the chickens started heading home to roost.

The investigators spent approximately a week onboard and determined that, as the saying goes, "Mistakes were made and people were hurt." There were findings of fault made, but before they could be acted upon, we had transited through the area of operations and into the region of the next fleet commander...the one to whom we were assigned when in home port. Our first stop for fuel found our Commodore waiting for us on the pier, and he did not look amused. This initial impression was supported when he announced he would be transiting back to Pearl Harbor with us. On the way, he interviewed every person on the ship. In the end, well over a dozen enlisted sailors faced non-judicial punishment for both the crossing the line ceremony and the other events that had led up to it. Officers' legal proceedings are traditionally not shared beyond the officer facing charges and the officer with authority to punish, with the exception of high visibility cases which draw the attention of the press. As such, I never knew if there were penalties handed down for the Shellback Department Heads onboard who may or may not have seen or participated, or the Executive Officer or Commanding Officer.

This twenty-five year old example, and it's fairly long-winded description, illustrate several timeless points. First, behavior has consequences. Sailors who went beyond what was usual and accepted were subsequently punished for their behavior. Officers who chose not to participate in a positive manner had their leadership called into question at a minimum. I made several attempts to look up the ship's senior leadership a few years later, and at the department head level and above, they all appear to have left the service within five years after the event. That may be unrelated to the events on the ship's deployment, but it is unusual and curious.

At a personal level, as I mentioned before, I had a pretty good time during the ceremony, but it has bothered me for years that I antagonized both the Commanding Officer and some of the sailors, who may very well have been subsequently punished for their response. This may be an over-inflation of my centrality in the story, as they would likely have done the same if I were a thousand miles away, but it bugs me nonetheless. Second, and this is more important, someone is always watching. When leaders hint at excessive behavior being OK, they have tacitly approved it. When a leader chooses not to stop it, or at least check on it, he or she is the one who is in the end responsible.

Every one of the men I've discussed in this story is an otherwise stand-up human being, and there are no villains. The sailors who beat me black and blue are friends of mine on Facebook, and I routinely exchange quips with them. If one of them needed a kidney I'd drive four states to their hospital today for the test. The Commanding Officer has been out of the service for over two decades and recently found some old pictures of the deployment. He was considerate enough to take the time to look many of us up and forward us the shots. People make mistakes and have to live with the consequences. Don't be afraid to draw the line when necessary. You'll be sorry if you don't."

Chapter Twelve
Setting the Example

Although this was originally listed as a bullet in the middle of the Navy's list, in my mind it is the seminal leadership principle, as all the other principles naturally flow from it. If leaders did what they expected of their subordinates on a daily basis, the world would flow smoothly. Unfortunately, that isn't the case. Perhaps it is because we have become jaded by the media's coverage of the political process, but people are far more attuned to what leaders do than what they say. As a leader, you can tell people to "do as I say, not as I do," and it will work for a little while, especially in a hierarchical organization like the military, but if you want them to understand a behavior is important, you have to show them by modeling that behavior yourself.

This is something kids notice from an early age. Teachers may say things like "keep off the grass" or "it's important to meet your commitments (do your homework)," but if they don't emulate the behaviors they are trying to impart to the children, the kids immediately realize the behaviors aren't important after all. It only takes one time when the teacher doesn't grade a test as promised, and students begin to doubt the teacher.

Having learned this discernment process at the lowest levels and the youngest ages, it is no surprise it exists full-blown at the adult level. I've saved the following story for this principle, because I feel strongly about it...

Story #33. "The Little Command that Couldn't"

"During my time with the military, I was attached to a unit that had the (bad) luck of having to work closely with the most dysfunctional unit I have ever seen. A friend of mine had worse luck; he was a member of the Training Department of that unit. Because of my close association with both the command and a few of its members, I had a front row seat to watch the worst example of leadership I have ever seen.

My friend arrived at his command shortly before I found my way to my own. He had always been a hard charger in his career, attaining every qualification he could in minimum time. He was a rising star.

Until he hit "The Command."

It was apparent the command had a problem the first time I visited it; there was no one there. It was a ghost town. That could have been expected for a Saturday or maybe even a late Friday afternoon, but not for a Tuesday afternoon at 1300 (1:00 p.m.). It was a large command; there should have been a lot of people around working on the command's equipment.

There weren't.

Deciding there must be some sort of command function that occupied all the command's personnel, I didn't give it another thought.

Until I went by the command again at 1130 (11:30 a.m.) on a Friday morning, and the parking lot was still empty.

I happened to hold the door for the commanding officer as I walked into the building (he was walking out.) I asked him how he was, and he said, "Great; it's Friday before noon, and I'm leaving for the week."

Now, I've known a lot of commands in my three-decade association with the Navy, and I've seen commands released early for the weekend many times. Sometimes they just got back from deployment and were receiving some well-deserved stand-down time. Others times, it was because the command met a goal set for it.

Not this time.

Even though the established working hours were from 0700-1600 (7:00-4:00), people left before noon every Friday, and several hours early on the other days of the week. As I came to realize the longer I was associated with the command, it wasn't because they had won a reward; it was because there was a complete lack of work ethic in the command. They were lazy. It started at the top and trickled down through the officers to the chiefs to the enlisted men and women. They did the minimum work required, jealously guarding their off-duty time, which seemed to come earlier and earlier every week. Most days, they couldn't even be bothered to flush toilets in the bathroom or pick up garbage in the passageways; they were that lazy.

Not surprisingly, the command's equipment wasn't in the best shape (because no one was ever there to work on it), a fact I saw borne out several times on the infrequent occasions when training events occurred. If the command scheduled four of their craft for flights, they would rarely end up with more than two available, leading to missed events and numerous missed training opportunities. This led to a vicious cycle of negative morale; because the equipment was broken, no one got trained, and because no one got trained, they didn't feel like doing the work to get people ready to be trained. Their attitudes were lousy. After several repetitions of this cycle, it had been reinforced to the point that most stopped caring.

I could see the same thing happening to my friend, who was in a middle management position. At the beginning, he was gung-ho to fix the process. He knew what needed to be done (it was pretty apparent), and was going to do it. After a few months, he was no longer going to fix the process; instead, he told me he was going to "try to guide it in the right direction to where it could be fixed." After six months, he rarely left his office (during the few hours he was there). He had been coopted by the process. He was no longer part of the solution; he had banged his head on the wall so long he finally succumbed to the overall malaise of the command...and now he was part of the problem.

The command needed Leadership (all of it) to step in and "right the ship," but this never occurred. The commanding officer was very pleasant on our infrequent meetings; however, I never saw the CO walking around inside the command. I'm not sure the CO was even aware the command *had* a problem. Like their commanding officer, the same could be said for the majority of the officers, as most of them stayed in their offices and were very disconnected from the troops (who were quite vocal about their displeasure). Unlike every other command I was ever in or around, the junior sailors were focused solely on going home, not on mission accomplishment.

How did they get this way? Their officers and chief petty officers set the example that this was an accepted behavior by modeling it for them, day in and day out. It started at the top. Remember that working hours started at 7:00? The CO got there every day between 8:20 and 8:30, setting the example that working hours weren't important.

If it wasn't important for the CO to be there for work, it didn't take long before the work itself became meaningless. That was what happened throughout the entire command. The CO set the example it was okay not to care…and everyone followed his example. One month, they had a mission completion rate of 16%; I'm only surprised it was that high.

Although there's nothing more instructive than a bad example, there's also nothing more *destructive* than a CO who sets a bad example."

Of all the leadership principles espoused by the military, this one is the most applicable to the civilian world, as it is based on human nature and goes well beyond the boundaries of the military services. As children, we learn that people can "talk the talk," but it's only important if they also "walk the walk." In other words, what people say isn't as important as what they do. We're taught throughout our lives that if we want to know what's important, we should watch what a leader does, not what he or she says. The fact that so many of our political leaders tell us what we should do instead of setting the example themselves is a major reason why the general public has lost faith in the political process; the leaders refuse to set the example.

And subordinates *will* follow the example set by their leaders, for better…or worse.

Chapter Thirteen
Leading by Example

As was discussed in Chapter Two, one of the most heavily-debated leadership questions is whether "Leadership" is an art or a science. Is leadership a trait that someone naturally has, or is it something that can be studied and taught? I believe the answer to this question is "yes; it is both an art *and* a science." There are people who are born with more charisma than others, and who come to leadership more easily. There are, however, key behaviors that can be taught and emulated, which will help people become better leaders, otherwise this guide is nothing more than a bedtime storybook of leaders acting badly. The point of this book is to show where leaders went wrong. As George Santayana noted in his book, *The Life of Reason*, 'Those who cannot remember the past are condemned to repeat it.' It is far better to learn from the mistakes of others than have to learn from your own.

When leaders say one thing but do another, they erode trust, a critical element of productive leadership; therefore, good leaders *must* lead by example. By aligning actions with speech, they become leaders whom subordinates want to follow and emulate. There is an enormous number of ways you can lead by example. If you only remember and model these six, though, you will be well on your way to being the leader your subordinates (and seniors, too, for that matter) want you to be.

1. Be accountable. Take responsibility for the unit. If something goes wrong, what could you have done to better train and equip your subordinates? Not only do you lose credibility when you blame others, it also keeps your subordinates on the defensive, and they won't grow when they're being blamed. Remember the maxim, "Praise in public; scold in private." Be accountable for your unit when in public and handle any failures behind closed doors.

2. Be future-focused. Until someone builds a time machine, you can't change anything in the past. When failures occur, which they will, look for the cause and, more importantly, what can be done to fix it in the future. When I found my secretary had been stealing money from the organization I was in charge of, I didn't try to cover it up; instead, I acknowledged the failure, and I told the stakeholders what I intended to do to ensure it didn't happen again. People appreciate the honest approach. Failures that are covered up only fester until they are ultimately exposed…and then are far worse for it.

3. Be competent. You may not be the best at what your unit does when you take over as the leader; you may not even know everything the unit does. That's okay. Listen. Ask questions. Seek to understand. You'll receive valuable insights and will set a tone that encourages healthy dialogue. You will also make your subordinates feel like you appreciate them and value their judgment, which you hopefully do!

4. Be honest. Even children can tell when they're being lied to, and they don't appreciate it. Say what you mean and mean what you say, and then *make your actions match your words!* When subordinates see you are honest and follow through on what you say you'll do, they

learn to trust you. There will be times you'll have to ask them to do things that are unpleasant (stay late, work the weekend, take that hill); if they trust you, they will do what you ask them to…on your word alone.

5. Be a delegator. Give your subordinates a chance to excel and support them in their efforts. With delegation comes the responsibility to equip your subordinates with the tools and training they need to successfully accomplish the task. Set them up for success, not failure! When someone is out of the office, is the next person in seniority able to step up and fill in? They should be, even for you, without missing a beat.

6. Be balanced. Work is important, but so is play. Take time to exercise and do things outside of the job. It takes a balanced leader, mentally and physically, to create a successful team. Anything you require of your subordinates, you have to do first, whether that is coming to work on time, learning to operate a new piece of gear or staying physically fit.

Subordinates judge you every moment of every day, looking to see if you are worthy of their trust. They want to know a number of things. Do you care about them and have their best interests at heart (remember the convoy commander job?) Can they trust you? Will you follow through and deliver what you promise?

The two biggest qualities they're looking for are competence and compassion. Competence is all about knowing what you're doing and being able to fulfill the duties of the position. While you're doing those things, will you take care of your people? If so, you're compassionate. You don't have to be cold and distant to be a leader. You can choose to be liked.

The most effective leaders I've known are the ones who empathize (or, if they can't empathize, at least try to sympathize) with the people they are charged to lead, no matter whether it is on the battlefield or in the factory. The military is blessed with a culture of leadership where junior officers and enlisted sailors are taught leadership from early on in their careers. In most cases, people "get it" and understand what they're being taught; it is only in a minority of cases where individual leaders choose to treat their commands as their own personal playthings.

Most of the officers I have served with and served under have been good leaders, which is due in large part to the principles of leadership the military follows. The principles in this book are solid principles that are universally applicable, whether you are a leader in the military or in civilian life.

If anything, these principles are even more important outside the military, as civilian life exists without the structures the military has to prevent the abuse of power (such as transferring people every 2-3 years). If civilian leaders follow these principles, they will be successful. If not…well, you saw the stories; you can guess how it turns out.

Speaking of stories, this has been a book of stories, and there's one more to illustrate this point…

Story #34. "The Last Slide"

"A friend of mine taught a leadership class for the Navy. By Friday afternoon, people would get restless, and it would be obvious the students were ready to leave for the weekend. He would always end by saying, "Okay, if I could have your attention, we have just one more, short lesson before we all leave." Sufficiently focused, the class would grow quiet, and he would say, "Here we go." He'd turn the projector back on and bring up the first slide.

It said simply, "I care."

As everyone stared at the slide, trying to figure out what it meant, he would say simply, "Just kidding; nobody teaches that.""

Don't be that person. It doesn't matter what industry you're in; if you care about the people you lead, they will know it and reciprocate by doing their best for you. Set the example, and your subordinates will follow. Be the leader your subordinates deserve.

ABOUT THE AUTHOR

A bestselling Science Fiction/Fantasy author and speaker, Chris Kennedy is a former naval aviator with over 3,000 hours flying attack and reconnaissance aircraft, an elementary school principal and an instructional systems designer for the Navy. Chris is also a member of the Science Fiction and Fantasy Writers of America and the Society of Children's Book Writers and Illustrators. Called "fantastic" and "a great speaker," he has coached hundreds of beginning authors and budding novelists on how to self-publish their stories at a variety of conferences, conventions and writing guild presentations.

Chris' full length novels include the "Occupied Seattle" military fiction duology, "The Theogony" and "Codex Regius" science fiction trilogies, and the "War for Dominance" fantasy trilogy. He is also the author of the #1 self-help book, "Self-Publishing for Profit."

Titles by Chris Kennedy:

"Red Tide: The Chinese Invasion of Seattle" – Available Now

"Occupied Seattle" – Available Now

"Janissaries: Book One of the Theogony" – Available Now

"When the Gods Aren't Gods: Book Two of the Theogony"
– Available Now

"Terra Stands Alone: Book Three of the Theogony"
– Available Now

"The Search for Gram: Book One of the Codex Regius"
– Available Now

"Beyond the Shroud of the Universe: Book Two of the Codex
Regius" – Available Now

"Self-Publishing for Profit" – Available Now

* * * * *

Connect with Chris Kennedy Online:
Facebook: https://www.facebook.com/chriskennedypublishing.biz
Blog: http://chriskennedypublishing.com/
Want to be immortalized in a future book?
Join the Red Shirt List on the blog!

www.ingramcontent.com/pod-product-compliance
Lightning Source LLC
Chambersburg PA
CBHW071702210326
41597CB00017B/2295